Springer Series in Materia

Volume 319

The Springer Series in Materials Science covers the complete spectrum of materials research and technology, including fundamental principles, physical properties, materials theory and design. Recognizing the increasing importance of materials science in future device technologies, the book titles in this series reflect the state-of-the-art in understanding and controlling the structure and properties of all important classes of materials.

More information about this series at http://www.springer.com/series/856

Yanjie Su

High-Performance Carbon-Based Optoelectronic Nanodevices

Springer

Yanjie Su
Shanghai Jiao Tong University
Shanghai, China

ISSN 0933-033X ISSN 2196-2812 (electronic)
Springer Series in Materials Science
ISBN 978-981-16-5499-2 ISBN 978-981-16-5497-8 (eBook)
https://doi.org/10.1007/978-981-16-5497-8

This Springer imprint is published by the registered company Springer Nature Singapore Pte Ltd.
The registered company address is: 152 Beach Road, #21-01/04 Gateway East, Singapore 189721, Singapore

Preface

Different sp^n hybridizations of carbon atoms result in various carbon allotropes with zero-, one-, and two-dimensional nanostructures, which show special structure-dependent physical and chemical properties. Especially, 1D single-walled carbon nanotubes (SWCNTs) can behave as metallic or semiconducting depending on the diameter and chirality, and the interband transitions between van Hove singularities determine the optical absorption and corresponding photoelectric properties of semiconducting SWCNTs (sc-SWCNTs). Consequently, superior intrinsic mobility, high absorption coefficient, and diameter-dependent bandgap feature of sc-SWCNTs open up exciting new possibilities for next-generation optoelectronic devices at room temperature. Different from SWCNTs, single atomic-layer graphene is a zero bandgap semimetal with ultrahigh carrier mobility and ultrafast carrier dynamics. Although monolayer graphene has a low optical absorption ($\sim$2.3%): unique electronic structures with linear dispersion make graphene absorb photons with different energies in the wide region from UV to THz range, enabling graphene to be a great promising potential for applications in photonics, electronics and optoelectronics. Therefore, high-performance carbon-based optoelectronic nanodevices have attracted significant attentions since the discovery of SWCNTs and graphene.

As we know, the performance of optoelectronic nanodevices is mainly affected by the purity and chirality of sc-SWCNTs or the quality of graphene. Especially, the existence of metallic/semiconducting SWCNTs and sc-SWCNTs with mixed chiralities strongly hinters the optoelectronic performance of SWCNT-based nanodevices owing to high recombination probability in SWCNT network or film. In the past few years, with the development of synthesis techniques and in-depth understanding, the fundamental properties of carbon nanostructures, ultrahigh purity sc-SWCNTs, or single-chirality SWCNTs with specific bandgap have been successfully synthesized by direct chirality-controlled growth or subsequent separation, and wafer-scale graphene has also been exploited for the application of electronic and optoelectronic nanodevices. Together with novel device designs and fabrication techniques, SWCNTs and graphene are prospective candidates for fabricating next-generation optoelectronic nanodevices in the practical applications.

The main purpose of this book is to introduce the research progress about optoelectronic nanodevices based on SWCNTs, graphene, and their heterojunctions in the past few years, and the corresponding future prospects have also been given. In order to help the readers understand more systematically carbon-based optoelectronic nanodevices, this book firstly introduces the basic physic features, controlled growth methods, and common-used characterizations of SWCNTs and graphene. Meanwhile, the basic physical mechanism and figures-of-merit of devices have been introduced before summarizing the research progress about carbon-based photodetectors and solar cells. Finally, all-carbon optoelectronic devices, in which both electrodes and active materials are composed of different carbon nanostructures, have been discussed, and the emerging trends have also been highlighted. We strongly believe this book can provide comprehensive information for academic and industrial researchers in the related fields. This monograph also serves as a professional reference book for physicists, material scientists, chemists, and engineers working on carbon nanomaterials and related nanodevices.

Shanghai, China Yanjie Su

Acknowledgements I would like to express my gratitude to Prof. Yafei Zhang for his encouragements and supports in the past decades. I would also like to thank all my graduate students who have participated in editing and collating this monograph. I also acknowledge the support from the National Natural Science Foundation of China (No. 61974089) and the Shanghai Natural Science Foundation (No. 19ZR1426900) for the research work in the book.

Contents

1 Introduction of Carbon Nanostructures 1
1.1 Carbon and Carbon Allotropes 1
1.2 Low-Dimensional Carbon Nanomaterials 4
1.2.1 Zero-Dimensional Carbon Nanomaterials 4
1.2.2 One-Dimensional Carbon Nanomaterials 5
1.2.3 Two-Dimensional Carbon Nanomaterials 6
1.3 Applications of Carbon Nanotubes and Graphene 8
1.3.1 Electronic Nanodevices 8
1.3.2 Optoelectronic Devices 10
1.3.3 Photovoltaic Devices 13
1.3.4 Chemical Sensing Nanodevices 14
1.3.5 Energy Storage Devices 16
1.3.6 Thermal Management 18
1.3.7 Biology 19
1.4 Future Perspective of Carbon Nanotubes and Graphene 20
References 21

2 Basic Physics of Carbon Nanotubes and Graphene 27
2.1 Basic Structure of Graphene and SWCNT 27
2.2 Band Structure and Electronic Transport of Graphene 29
2.3 Band Structure and Electronic Transport of SWCNTs 31
2.4 Optical Properties of SWCNTs 37
2.5 Optical Properties of Graphene 38
References 39

3 Controlled Growths of Carbon Nanotubes and Graphene 41
3.1 Arc Discharge Synthesis of SWCNTs 41
3.2 Laser Ablation Synthesis of SWCNTs 45
3.3 The CVD Synthesis of SWCNTs 46
3.3.1 The Floating-Catalyst CVD of SWCNT Powders 47

3.3.2 The Catalyst-Supported CVD of SWCNTs 48
3.4 Catalytic CVD Growth of Graphene on Metal Substrates 53
3.5 Catalytic CVD Growth of Graphene on Nonmetal Substrates 55
3.6 Catalytic-Free CVD Growth of Graphene on Dielectric and Insulating Substrates 56
3.7 Summary and Outlook 58
References 58

4 Characterizations of Carbon Nanotubes and Graphene 65
4.1 Introduction 65
4.2 Raman Spectroscopy 66
4.2.1 SWCNTs 66
4.2.2 Graphene 70
4.2.3 Charge Transfer Analysis of SWCNTs and Graphene 73
4.3 Electron Microscopy 75
4.3.1 SWCNTs 75
4.3.2 Graphene 78
4.4 UV-vis-NIR Absorption Spectroscopy 79
4.5 Photoluminescence Spectroscopy 82
4.6 Other Characterization Techniques 84
References 85

5 Carbon-Based Heterojunction Broadband Photodetectors 91
5.1 Introduction 91
5.2 Physical Mechanism and Key Parameters of Photodetectors 92
5.2.1 Physical Mechanism 92
5.2.2 Key Parameters 94
5.3 CNT-Based Photodetectors 95
5.3.1 High-Performance SWCNT Photodetectors 96
5.3.2 CNT/Bulk Semiconductor Heterojunction Photodetectors 101
5.3.3 CNT/Nanomaterial Heterojunction Photodetectors 105
5.3.4 SWCNT/MoS_2 vdW Heterojunction Photodetectors 107
5.4 Graphene-Based Photodetectors 109
5.4.1 High-Performance Graphene Photodetectors 110
5.4.2 Graphene/Bulk Semiconductor Heterojunction Photodetectors 112
5.4.3 Graphene/QD Heterojunction Photodetectors 113
5.4.4 Graphene/2D Semiconductor vdW Heterojunction Photodetectors 116
5.5 Summary and Outlook 121
References 122

6 All-Carbon van der Waals Heterojunction Photodetectors 131
6.1 Introduction 131
6.2 The Stacking Methods of All-Carbon vdW Heterostructures 132
6.3 SWCNT/C_{60} vdW Heterostructures 134
6.4 Graghene/C_{60} vdW Heterostructures 135
6.5 Graghene/Carbon QD vdW Heterostructures 137
6.6 Graphene/SWCNTs vdW Heterostructures 138
6.6.1 Charge Transfer Between Graphene and SWCNTs 139
6.6.2 Graphene/SWCNT vdW Heterojunction Photodetectors . . . 139
6.7 Summary and Outlook 142
References 144

7 Carbon Nanotube/semiconductor van der Waals Heterojunction Solar Cells 149
7.1 Introduction 149
7.2 Work Mechanism and Key Parameters 150
7.2.1 PN and Schottky Heterojunctions 150
7.2.2 Key Parameters of Solar Cells 152
7.3 DWCNT or MWCNT/Si vdW Heterojunction Solar Cells 153
7.3.1 Improving the Conductivity of the CNT Films 153
7.3.2 Changing the Fermi Level of MWCNT Films 155
7.4 SWCNT/Si vdW Heterojunction Solar Cells 157
7.4.1 Improving the Structure and Properties of SWCNT Films 157
7.4.2 Optimizing the Fermi Level of SWCNT Films by Chemical Doping 160
7.4.3 Improving the Hole Transportation of SWCNT Films 162
7.4.4 Introducing Insulating Interfacial Layers 163
7.5 SWCNT/GaAs vdW Heterojunction Solar Cells 164
7.6 Summary and Outlook 166
References 167

8 Toward All-Carbon Hybrid Solar Cells 171
8.1 Introduction 171
8.2 Excitation Transfer Dynamics of Sc-SWCNTs and Sc-SWCNT/Fullerenes 172
8.3 Bandgap Structure Limits of SWCNT/Fullerenes Hybrids 173
8.4 All-Carbon Nanohybrids as Active Layer 174
8.4.1 All-Carbon Bulk Heterojunctions 175
8.4.2 All-Carbon Planar Heterojunctions 176
8.5 All-Carbon Solar Cells 180
8.6 Summary and Outlook 181
References 184

Index 187

Abbreviations

0D	Zero-dimensional
1D	One-dimensional
2D	Two-dimensional
AFM	Atomic force microscopy
BHJ	Bulk heterojunctions
BP	Black phosphorus
BWF	Breit–Wigner–Fano
CNT	Carbon nanotube
CQDs	Carbon quantum dots
DIO	1, 8-diiodooctane
DOS	Density of states
DWCNT	Double-walled carbon nanotube
ED	Electron diffraction
EES	Electrochemical energy storage
EQE	External quantum efficiency
ERS	Electronic Raman scattering
FF	Fill factor
FFT	Fast Fourier transformation
FWHM	Full width at half-maximum
GO	Graphene oxide
GQDs	Graphene quantum dots
HiPco	High-pressure carbon monoxide
IQE	Internal quantum efficiency
LIBs	Li-ion batteries
LSPR	Localized surface plasmon resonance
MEG	Multiple exciton generation
MIR	Mid-infrared
MOCVD	Metal-organic chemical vapor deposition
MPA	Multiphoton absorption
MWCNT	Multi-walled carbon nanotube

NIR	Near-infrared
OA	Triethyloxonium hexachloroantimonate
OPV	Organic photovoltaics
PANI	Polyaniline
PCBM	[6,6]-phenyl-C61-butyric acid methyl ester
PCE	Power conversion efficiency
PEDOT	Poly(3, 4-ethylenedioxythiophene)
PEDOT:PSS	Poly(3, 4-ethylene dioxythiophene):poly(styrenesulfonate)
PEI	Polyethylenimine
PL	Photoluminescence
PLD	Pulsed laser deposition
PLE	PL excitation
PMMA	Poly(methyl methacrylate)
PPy	Polypyrrole
PTE	Photothermoelectric
QDs	Quantum dots
RBM	Radial breathing mode
RF	Radio frequency
rGO	Reduced graphene oxide
rr-P3DDT	Regioregular poly(3-dodecylthiophene-2,5-diyl)
SAED	Selected area electron diffraction
SCs	Supercapacitors
SEM	Scanning electron microscopy
SOI	Silicon-on-insulator
STM	Scanning tunneling microscopy
SWCNT	Single-walled carbon nanotube
TCNE	Tetracyanoethylene
TCNQ	Tetracyanoquinodimethane
TCR	Temperature coefficient of resistance
TEM	Transmission electron microscopy
TMD	Transition metal dichalcogenides
TTF	Tetrathiafulvalene
UPS	Ultraviolet photoelectron spectroscopy
vdW	van der Waals

Chapter 1
Introduction of Carbon Nanostructures

1.1 Carbon and Carbon Allotropes

As one of the most important Group IV elements, the atomic number of carbon is 6 with the atomic weight of 12.011, which is considered as the backbone element of all organic materials and living organisms on earth due to multiple possibilities of chemical bonding. Meanwhile, it also exists widely in various complex compounds in the universe and plays an important role in the evolution of the universe. Actually, the utilization and exploitation of carbonaceous matters and carbon materials have always played a great role in driving social progress and human development. Especially, the coal and coke were first used as a fuel and then to make copper and iron metals by reducing metal ores. Until now, the coal and coke are still the most used carbon materials in modern society. With the development of industrial technologies, other physical properties of carbon materials are exploited and widely used, such as the electrical conductivity, high-temperature resistance, high strength, lubricity, and so on. For instance, high-quality graphite has been widely used as electrodes in metal smelting industry and as a neutron moderator and reflector material in nuclear reactor. Diamond as the toughest material known in nature can be used in cutting, grinding and drilling, while high thermal conductivity, high optical transparency and excellent electrical insulation makes it be widely used in the electronic industry. Carbon fiber exhibits lightweight, high strength, high modulus, high temperature resistance, corrosion tolerance and fatigue resistance properties, which has played a key role in producing high-performance composites mainly used in aerospace and aeronautical fields. On the other hand, activated carbon with large porosity has a broad range of applications in air purification, wastewater treatment, energy storage, antisepsis, medicine, and so on. Therefore, carbon materials are very closely related to the human life and production in modern society [1–3].

However, for a long time, graphite and diamond are the only allotropes of carbon until the Nobel Prize-winning discovery of fullerene (C_{60}) in 1985. Since then,

Y. Su, *High-Performance Carbon-Based Optoelectronic Nanodevices*,
Springer Series in Materials Science 319,
https://doi.org/10.1007/978-981-16-5497-8_1

nanomaterials and nanotechnology have started to play an increasing role in promoting the progress of Sci-Tech and economic development. The subsequent discovery of carbon nanotubes (CNTs) in 1991 has attracted continuous and extensive attentions in various fields due to unique atomic structures and structural-depended properties. In particular, single-walled CNTs (SWCNTs) have been demonstrated to fabricate higher performance electronic, optoelectronic, photovoltaic, and sensing nanodevices. The first preparation of graphene in 2004 made it have a great potential to the application in photonics, optoelectronics, transparency electronics, and so on. Together with new carbon allotropes (atomic carbon chains and graphdiyne): the properties and application fields of carbon materials will be further expanded in 21st century. Therefore, the development of these carbon allotropes is promoting mankind into a new era of carbon, [3] in which carbon nanomaterials and nanodevices will be widely used in the next-generation information processing technology, and the performance of carbon-based integrated circuits will be much higher than those of silicon [4, 5].

Actually, the unique properties and abundant types of carbon materials are essentially determined by the nature of electronic structure and the hybridization states of carbon atoms. A carbon atom contains six electrons with the following electron configuration: $(1s)^2(2s)^2(2p)^2$. The small energy difference between the 2s- and the 2p- orbital electrons makes it easily to excite one electron from the 2s state into the 2p-state, resulting in the formation of sp^n-hybrid orbital (n = 1, 2, 3) by mixing 2s- and the 2p- orbitals. Consequently, carbon atoms can form more valence bond configurations, various atomic arrangement structures and allotropes, while none of the other elements have this ability. In the sp^n-hybridization, the (n + 1) electrons in hybrid orbitals form a σ-bond, while additional unhybridized [4 − (n + 1)] electrons in 2p- orbitals form a π-bond. Obviously, there are only σ-orbital electrons in sp^3-hybridized materials (eg. Diamond): which involve the C–C bonding with a large binding energy and distribute in the direction of C–C binding axis. However, for most carbon materials, π-orbital electrons are also involved and distribute perpendicular to the C–C bonding axis, which can move freely in the nonlocal conjugated system formed inside and on the surface of the crystal and molecules, forming a π-orbital electron cloud in sp^2- and sp- hybridized materials. More importantly, σ-orbital electrons are sensitive to the change of physical and chemical environments, making carbon materials exhibit a variety of unique properties.

The unique electron configuration and multiple chemical bonding possibilities enable the formation of a large number of carbon allotropes, such as diamond, graphite, and amorphous carbon. The Nobel Prize-winning discovery of fullerenes (C_{60}) in 1985 has inspired remarkably research interest in carbon allotropes and carbon nanomaterials, and the subsequent discoveries of SWCNTs and graphene open up exciting new possibilities for carbon era. Nowadays, different carbon allotropes have been discovered and synthesized, and their electron orbital hybridizations can be summarized in Table 1.1. Except for diamond, unique

Table 1.1 Carbon allotropes and their electron hybridization types

/	sp^3	sp^2	sp	$sp^2 + sp^3$	$sp + sp^2 + sp^3$
0D	Nano-diamond	Graphene quantum dots	/	Fullerenes Carbon dots Carbon onions	Carbon black
1D	/	Graphene nanoribbons	Carbyne C-atom chains	CNTs Carbon nanofibers	/
2D	/	Graphene Graphene oxide	/	/	Graphdiyne
3D	Diamond	Graphite	/	Carbon fibers	Carbon black

physical and chemical properties of carbon materials are closely related to the existence of π-orbital electron cloud, such as ultrahigh carrier mobility, ballistic electron transport, excellent optoelectronic properties, and so on. For example, graphene is a typical sp^2-hybridized crystal structure, in which the planar sp^2-hybrid orbitals arrange carbon atoms into six-carbon rings through the σ-bonds with a strong binding energy, while π-orbital electron can move freely on the surface and exhibits an ultrahigh mobility in excess of 200,000 $cm^2\ V^{-1}\ s^{-1}$ due to its small effective mass (0.056 m_0) [6]. More importantly, the valence and conduction band of graphene consist of the bonding and antibonding π- and π*-orbital electrons, which meet at the K-points where the density of states (DOS) is found to be zero at the Fermi level. The unique electron structures enable graphene as zero-bandgap semiconductor exhibit many unexpected electronic and optoelectronic properties. However, when graphene sheet is rolled up to form SWCNT along a certain direction, the distribution of π-orbital electrons will be changed and the sp^2-hybrid orbitals will be deformed owing to the existence of curvature and helical angle in quasi one-dimensional structures, which are basically determined by π-orbital electrons. Consequently, except for ultrahigh carrier mobility similar to graphene, SWCNTs can behave semiconducting or metallic according to the arrangement of carbon atoms, and the physical and chemical properties are also remarkably depended on the diameter and helical angle. Especially, the one-dimensional nature produces van Hove singularities in the DOS of SWCNTs, and the diameter-depended exciton binding energy in SWCNTs is higher than typical bulk semiconductors due to the quantum size effect, resulting in a large radiative lifetime up to 100 ns and a fluorescence lifetime up to 100 ps at room temperature [7]. In fullerenes (e.g., C_{60}): π-orbital electron cloud is formed inside and outside the surface and π-electrons exhibits a discrete distribution in energy. As a result, the fullerenes exhibit typical n-type semiconducting properties with band gap of about 2.5 eV for individual molecules and about 1.5 eV for fullerene crystal [8]. Moreover, the band gap is inversely proportional with the diameter of fullerenes, which is similar to the band gap of SWCNTs.

1.2 Low-Dimensional Carbon Nanomaterials

Since the discovery of fullerenes, carbon nanostructures have attracted increasing research attentions from academia and industry in the past thirty years, Nowadays, various carbon nanomaterials with zero-, one- and two-dimensional structures have been reported, which can exhibit semiconducting, metallic or insulating, extremely dark or highly transparent, variable thermal conductivity due to the difference of atomic structures. In this section, we will mainly introduce several typical carbon nanomaterials, which consist primarily of sp^2 carbon atoms arranged in a hexagonal lattice.

1.2.1 Zero-Dimensional Carbon Nanomaterials

Fullerenes (or buckyballs) a kind of zero-dimensional (0D) closed cage molecules consisting of a spherical network of structurally equivalent sp^2-hybridized carbon atoms, which are composed of 12 pentagons and hexagons determined by the Euler's rule. C_{60} molecular is one of the most stable and abundant representative of the fullerene family, and it was discovered at 1985 and for the first time isolated in a pure state in 1990 [9]. Due to high curvatures and unique π-orbital electrons, the fullerenes and their derivatives exhibit unusual electronic, magnetic and chemical properties. Nowadays, thousands of fullerenes and their adducts with various remarkable properties have been synthesized and used in different fields, such as electrocatalysts, energy storage, photodetectors, solar cells, diagnostics, and biomedical sciences [10–12]. For example, fullerenes are known to be a typical n-type semiconductor as ideal electron acceptors, which are commonly used in organic photovoltaic (OPV) devices. Meanwhile, its adduct of C_{60}, phenyl-C61-butyric acid methyl ester (PCBM): has also been widely used in OPV and perovskite solar cells as an efficient electron transport layer [13, 14].

Except for fullerenes, fluorescent carbon-based quantum particles are an emerging class of 0D crystalline nanomaterials with less than 10 nm in diameter, which exhibit discrete quantized energy levels and size-depended optical properties due to the quantum confinement effects, which was accidentally discovered by examination of the residues obtained from the arc-synthesized SWCNTs [15]. These quantum particles can be simply divided into two types: carbon quantum dots (CQDs) and graphene quantum dots (GQDs). CQDs are spherical quantum particles with core–shell structures consisting of sp^2 and sp^3 carbon atoms, while GQDs are disk-like crystalline graphene with sizes smaller than 20 nm. Both 0D quantum materials can be doped by other elements or surface modified by chemical functionalization, tailoring their physical and chemical properties. Therefore, different from semiconductor QDs, carbon-based quantum particles exhibit excellent water dispersibility, strong and tunable photoluminescence, exceptional physicochemical properties, and biocompatibility, which have been considered to be promising materials for optical, electroanalytical and biomedical applications.

1.2.2 One-Dimensional Carbon Nanomaterials

CNTs are the most typical one-dimensional (1D) nanomaterials with nanometer diameter and ultrahigh length/diameter ratio, which were first observed from arc-discharge graphitic soot using transmission electron microscopy in 1991. Subsequently, SWCNTs had been successfully synthesized in large-scale using arc discharge techniques with Fe as catalysts in 1993. Structurally, SWCNTs can be considered as a cylinder formed by rolling up a monolayer graphene, that is, a graphene sheet is rolled up along the chiral vector C = na_1 + ma_2 (a_1 and a_2 are lattice unit vectors) from the origin point (0, 0) to the point (n, m): forming single-chirality SWCNTs with specific chiral indices (n, m), [16] as shown in Fig. 1.1. Importantly, the chiral indices (n, m) do not determine the structures (diameter and helical angle) of SWCNTs, but also dominate the physical and chemical properties, such as conductive type, electronic structures, optical absorption and chemical activity, and so on.

As we know, SWCNTs can be either metallic or semiconducting depending on their diameter and chirality, one third of SWCNTs are metallic and the others are

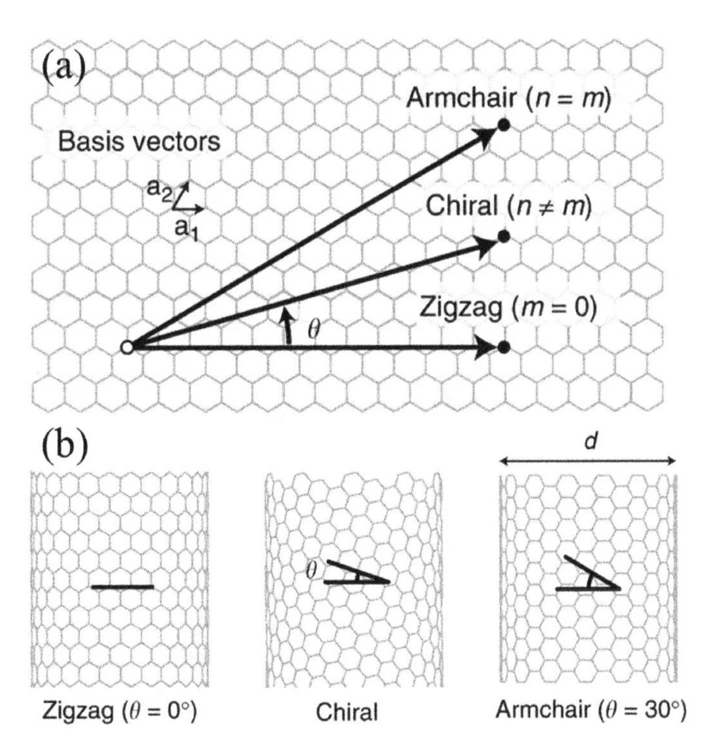

Fig. 1.1 **a** Schematic diagram of a graphene sheet with common chiral vectors. **b** Zigzag, chiral and armchair SWCNT with different chiral angles. Reproduced with permission [16]. Copyright 2019, Springer Nature

semiconducting. Especially, semiconducting SWCNTs (sc-SWCNTs) show a diameter-depended direct bandgap with 0.5–1.2 eV when the diameters range from 0.7 to 2.0 nm, [17] and can absorb light in a wide spectral range due to unique interband transitions (S_{11}, S_{22} and S_{33}) between van Hove singularities. Especially, strong optical absorptivity ($\alpha > 10^5$ cm^{-1}) enables sc-SWCNTs effectively collect the incident lights in films as thin as 100 nm. Coupling with ultrahigh carrier mobility as high as 10^5 cm^2 V^{-1} s^{-1} at room temperature, sc-SWCNTs have been considered to be a great promising material for next-generation integrated circuit, nanoelectronics, optoelectronics, sensing devices [18–20]. In addition, large specific surface area and physicochemical properties enable sc-SWCNTs to be used in biological and biomedical applications after biomodification [21]. While metallic tubes usually act as an efficient charge transport channel in nanodevices due to ultrahigh carrier mobility and current carrying capacity. Meanwhile, metallic SWCNT network can also be used as highly transparent electrodes for novel nanoelectronics, solar cells, and flexible nanodevices [22].

For multi-walled CNTs (MWCNTs) embedded by several SWCNTs, their physical properties are already quite different from SWCNTs due to the embedded structure of several SWCNTs, and the mechanical and electrical characteristics have been commonly used in the research and practical applications. For example, MWCNTs as matrix reinforcing agent has been commonly used to improve the mechanical performance of nanocomposites, including metal, polymers and biomaterials. Meanwhile, MWCNTs as conductive additive have been demonstrated to improve the discharge performance and capacity of lithium ion batteries, as well as high specific capacitance of supercapacitors [23, 24]. In addition, the outstanding intrinsic thermal conductivity makes MWCNTs to be good thermal interface material for thermal management.

1.2.3 Two-Dimensional Carbon Nanomaterials

Graphene is another important allotrope of carbon in the form of a two-dimensional (2D): atomic-scale, and stable hexagonal lattice, which is usually considered as a unit structure of graphite, CNT, and fullerene, as well as aromatic molecules with infinite size. Monolayer graphene composed of a closely packed single layer of carbon atoms was first mechanically exfoliated by Andre Geim and Konstantin Novoselov in 2004, [25] whose discovery won the Nobel Prize in Physics in 2010. Although the thickness of monolayer graphene is only 0.35 nm, the atomic connection is tough enough to endure the external force induced by twisting a lattice plane, avoiding the reconstruction of carbon atoms. Since a graphene sheet only consists of one atomic carbon layer, the planar density is as light as 0.77 mg/m^2. Meanwhile, one atomic thickness also makes graphene exhibit a very high transparency with 2.3% absorption for visible light. Owing to the existence of π-orbital electron cloud, graphene has a super electron mobility of 2×10^5 cm^2 V^{-1} s^{-1} at room temperature, which makes it to be ideal conducting material in transparent electronics, photoelectric devices. Combining with unique optical properties, graphene has exhibited a great promising potential in various fields. In addition, superhigh thermal conductivity more than 5000 W/mK at room temperature further broadens the application field of graphene [26].

The excellent physical and chemical properties have attracted huge increasing interesting from academia to industry since 2004, and different preparation techniques have been developed, including mechanical exfoliation, chemical exfoliation, CVD and epitaxial growth [27–30]. Among them, mechanical exfoliation method is commonly utilized to prepare high-quality single crystal graphene monolayers from bulk graphite, which is mainly used to study the intrinsic properties of graphene due to its low exfoliated efficiency and poor scalability. With the development of the preparation techniques, the CVD and chemical exfoliation methods gradually became the main method to prepare different graphene in large scale. Especially, CVD has been considered as the most effective approach to synthesize large-area and high-quality graphene films, and centimeter-scale monolayer graphene on a commercial copper foil has been demonstrated in 2009 [31]. Since then, numerous efforts have been made to synthesize large monolayer graphene films on different metal substrates under different growth conditions, including polycrystalline or single crystal Cu films, Ni film or 3D porous foam, and their alloy films or foils. More importantly, wafer-scale high quality graphene has been well explored on dielectric substrates, such as SiO_2, TiO_2, and glass, and so on. The exciting technique advances are significantly promoting the rapid development of graphene-based nanodevices and systems, especially in electronics and optoelectronics. In addition, the chemical exfoliation methods are the main technique to synthesize graphene at low cost based on modified Hummers methods, [28, 29, 32, 33] in which graphite is highly oxidized to graphene oxide (GO) and exfoliated into thin GO nanosheets with hydrophilic functional groups, followed the reduction of GO into reduced graphene oxide (rGO) via various reduction strategies. Obviously, there are a large number of defects in the rGO nanosheets, which are mainly used as conductive additives, sensing materials, composite enhancer, biomaterial templates, and so on [34–37].

Except for graphene, graphdiyne is another important 2D carbon allotrope, which is composed of sp- and sp^2-hybridized carbon atoms, which was first synthesized via a cross-coupling reaction on the surface of copper using hexaethynylbenzene in 2010 [38]. The new 2D nanostructures possesses excellent chemical stability, high electrical conductivity, uniformly distributed pores, and tunable electronic properties, facilitating the potential applications in the fields of catalysis, energy, sensor, electronics and optoelectronics [39]. Meanwhile, the carbon–carbon triple bonds in graphdiyne enable the formation of different graphdiyne derivatives by introducing adatoms (e.g., fluorine, hydrogen, or oxygen) [40]. With the efforts in the past ten years, great progresses have been made in both experiment and theory. However, many challenges remain in the emerging field, and it is still very necessary and crucial to identify and deeply understand the intrinsic physical properties of graphdiyne, and the controlled synthesis of graphdiyne with specific layer number and stacking styles is still a great challenge.

1.3 Applications of Carbon Nanotubes and Graphene

In the section, we will focus on the promising applications of SWCNTs and graphene in the different fields.

1.3.1 Electronic Nanodevices

Among emerging electronic nanomaterials, sc-SWCNTs have been demonstrated to the great promising candidates for next-generation nanoelectronics and ICs due to the following advantages: (1) The atomically smooth surface of sc-SWCNTs without dangling bonds facilitates a scattering-free ballistic transport of carriers in SWCNTs due to long mean free paths. As a result, SWCNTs possess exceptional field-effect mobility exceeding 10^5 cm^2 V^{-1} s^{-1} at room temperature due to the scattering phase-space restrictions. (2) Small capacitance (<0.05 aF/nm) of SWCNTs makes them have obvious advantages in fabricating high-speed low-power nanoelectronics due to minimal parasitic capacitance, efficient gate coupling and low switching power consumption. (3) The atomic thickness of sc-SWCNT channel and small diameter enable easier control of the channel current, making sc-SWCNTs possible to be used to fabricate carbon-based transistors smaller than 5 nm technologies. Moreover, SWCNT-based CMOS technology has been demonstrated to outperform silicon-based technologies by several technology generations (Fig. 1.2) [1, 41].

Nowadays, SWCNT CMOS FETs have exhibited outstanding the performance that compares favorably to the state-of-the-art Si MOS FETs. Especially, the doping-free approach preserves the perfect lattice of SWCNTs, thereby eliminating entirely doping-related performance fluctuations and performance limits at low and high temperature [41]. Moreover, the threshold voltage of the SWCNT FET can be controlled by using a gate metal with a suitable work function, enabling the SWCNT CMOS FETs with zero threshold voltage. In addition, the ballistic transport in SWCNT devices can significantly suppress the fabrication error-related effects and the random scattering in the channel. With the rapid development of SWCNT synthesis techniques and SWCNT-based nanoelectronics, the 16-bit microprocessor based on the RISC-V instruction set has been demonstrated using more than 14,000 SWCNT CMOS FETs [42]. Importantly, high-purity (better than 99.9999%) and high-density (100–200 SWCNTs/μm) aligned sc-SWCNT arrays have been successfully prepared on 4-inch silicon wafers with full wafer coverage, which can satisfy the fundamental requirement for large-scale fabrication of digital SWCNT ICs [43]. Recently, Peng group further demonstrated high performance radio-frequency (RF) transistors and amplifiers based on aligned high-density sc-SWCNT arrays ($\sim$120 SWCNTs/μm) with a 50 nm gate length. The current-gain and power-gain cutoff frequencies are up to 540 and 306 GHz,

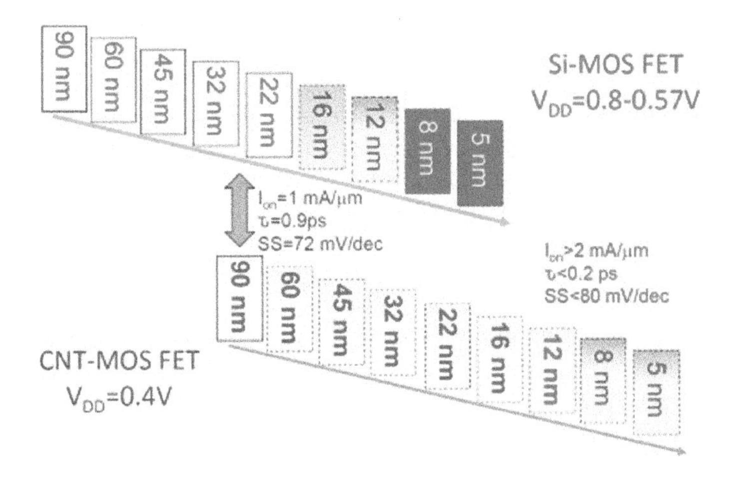

Fig. 1.2 Comparison of Si and SWCNT MOS FET roadmaps down to 5 nm transistor technology. Reproduced with permission [41]. Copyright 2014, Elsevier

respectively. The RF amplifiers could exhibit a high power gain of 23.2 dB and inherent linearity in the K-band (18 GHz) [44].

Different from sc-SWCNTs, the bandgap-free feature seriously limits the application prospects of graphene in the digital electronics, but it does not hinder the great advantages for RF electronic applications where the complete turn-off of a device is not essential. Therefore, graphene-based RF electronics have attracted more and more interesting in the past decades [45]. Firstly, the true single-atomic 2D layer structure makes graphene meet the requirement of the short gate length in high-speed RF devices. Secondly, room-temperature ultrahigh carrier mobility of graphene results in higher normalized transconductance parameter (up to 7 mS) than those in state-of-the-art Si MOSFETs and GaAs HEMTs. Along with small parasitic capacitance in graphene channel, high performance graphene RF electronics exhibits faster switching behavior and a very high cutoff frequency. Furthermore, the relative low contact resistance and high saturation velocity also allow graphene to be used to fabricate higher performance the RF electronics. With the development of controlled growth techniques of high-quality graphene, numerous research advances in graphene-based RF FETs have been constantly made, [45] and a cutoff frequency (f_T) of 100 GHz has exceeded the record set achieved by Si-based devices with the same gate length in 2011 [46]. And then, a record-high cutoff frequency of 427 GHz has been demonstrated in self-aligned graphene transistors, in which the unique structure of the transferred gate stacks was used to minimize the access resistance or parasitic capacitance [47]. Meanwhile, the maximum oscillation frequency exceeding 200 GHz of graphene FETs has been achieved by using a residue-free Au film as the sacrificial transfer and optimizing contact resistant [48].

1.3.2 Optoelectronic Devices

Unique band structures carrier mobility and ultrahigh enable SWCNTs and graphene great potential for the next-generation photoelectronic nanodevices. For example, van Hove singularities in the 1D DOS make sc-SWCNTs possess diameter-depended interband transitions due to its 1D confinement nature, which results in a broad optical absorption in the UV–visible-NIR range or a chirality-related photoluminescence. Moreover, the electronic transitions in SWCNTs are demonstrated to be sensitive to the polarized light owing to the specific angular momentum in sub-bandgaps. However, it is worth noting that the photoelectric properties of SWCNTs are mainly determined by the first interband transition (S_{11}) of sc-SWCNTs. For the sc-SWCNTs with a diameter of 0.7–2.0 nm, the S_{11} transition energy is in the range of 1.1–0.4 eV, well corresponding to the NIR spectrum range. Therefore, sc-SWCNTs are considered to be promising materials as novel NIR photodetectors and light emitters [1]. Since the quantum efficiency of light emission in SWCNTs is much lower than 10^{-3} and it has far away from possible practical applications, we mainly introduce the application of broadband photodetectors based on pure SWCNTs and their heterojunctions.

Compared with traditional bulk photodetectors, SWCNT-based photodetectors possess a series of remarkable advantages, such as fast light response, wide chiral-depended bandgap tunability, polarization sensitivity along the nanotube axis, room-temperature operation, good compatibility with conventional fabrication, and so on [20]. Thus, a large number of SWCNT-based photodetectors have been extensively studied in the past two decades, which can be simply classified into photoconductors, photodiodes, phototransistors and thermal detectors according to different detecting mechanisms [49]. From the viewpoint of active material consumption, it can be simply divided into three types: individual sc-SWCNT photodetector, aligned sc-SWCNT array photodetector and sc-SWCNT film photodetector. The earlier studies mainly focus on the investigation of the exciton dynamics and optoelectronic properties of individual sc-SWCNT when it was difficult to obtain pure sc-SWCNTs. The presence of metallic SWCNTs significantly reduce the non-radiative recombination time of charge carriers, dramatically suppressing the separation efficiency of photogenerated carriers in a SWCNT photodetector. To improve the separation efficiency of photogenerated excitons, different technique methods have been developed using asymmetric Schottky contacts, p/n junctions, and local chemical doping [49–51]. However, the weak optical absorption of individual SWCNT limits the further improvement of the detecting performance. With the development of separation technique and selective-synthesis of SWCNTs, the sc-SWCNT array or film has been extensively used to fabricate SWCNT-based photodetectors. For example, Liu et al. [52] demonstrated a self-powered SWCNT NIR photodetectors with asymmetric contacts, which exhibited a broadband response (785–2100 nm) and high detectivity (10^{11} Jones) at room temperature in voltage mode and under the light illumination, the largest voltage responsivity and detectivity are as high as 1.5×10^{8} V W^{-1} and

1.25×10^{11} Jones occurred at 1800 nm, respectively. More importantly, a wafer-scale photodetector array (150 × 150) has been fabricated with high uniformity on a 2-inch Si substrate, demonstrating a great potential for large-scale fabrication and imager applications. In addition, sc-SWCNTs possess a high Seebeck coefficient of $\sim$200 μV K^{-1}, which can be used as thermal detectors based on thermoelectric and bolometric effects, achieving a high response to broad electromagnetic spectrum from UV to THz region at room temperature [53, 54].

Except for pure SWCNTs as photoactive materials, SWCNT-based heterojunctions have also been extensively utilized to fabricate broadband photodetectors, in which the separation efficiency of photogenerated carriers can be enhanced significantly at the interfaces by suppressing the recombination rate in photoactive materials whether SWCNTs serve as photosensitive material or separation promoter. Up to now, a large number of SWCNT-based heterojunction photodetectors have been reported by combining different nanomaterials or bulk materials with mixed SWCNTs or sc-SWCNTs [55–57]. Among them, all-carbon van der Waals (vdW) heterojunctions are particularly attractive due to the combination of outstanding properties of SWCNTs and other carbon allotropes. For example, when the sc-SWCNTs combine with fullerenes (C_{60}, C_{70}, etc.) to form all-carbon p/n junctions, the electron transfer time at the interface is as fast as 120 fs, facilitating the development of high-speed response photodetectors. Using C_{60} as electron traps and sc-SWCNTs as an light absorbing/carrier generating layer, a sc-SWCNT/C_{60} vdW heterojunction planar phototransistor has been demonstrated, exhibiting a high responsivity nearly 200 A/W at V_{DS} of −1 V and Vg of 0 V under light illuminations [58]. This phototransistors can be further improved by thermal evaporating C_{60} layer on aligned sc-SWCNT arrays, which achieved an ultrahigh responsivity of 10^8 A/W under UV–visible illumination and 720 A/W toward 950 nm illumination [59]. Remarkably, the SWCNT/graphene vdW heterojunctions as the active materials are considered to have greater potential for high performance photodetectors, in which ultrahigh mobility, unique band structures, and strong regulatable electron coupling at the atomic interface can be fully utilized together. Liu et al. [60] demonstrated an all-carbon vdW heterojunction photodetectors by covering a monolayer continuous graphene on the surface of atomically thin SWCNTs, achieving a broad photoresponse across visible to NIR range (400–1550 nm). By replacing continuous graphene with separated graphene nanosheets, the photoelectric performance can be improved significantly, the device exhibited a broadband photoresponse with a high responsivity of $>3 \times 10^3$ A/W and a fast response speed of 44 μs, still representing the best result for SWCNT-based photodetectors so far [17]. It is believed that higher photoelectric performance could be achieved by increasing the network density of sc-SWCNTs and inhibiting the structural defects in sc-SWCNTs and graphene.

Different from SWCNTs, the gapless band structure and semi-metallic nature make graphene absorb the incident light in a wide spectrum from UV to THz region. Moreover, high mobility of graphene has been reported to yield ultrafast photodetection up to 40 GHz, and the intrinsic response speed is expected to be as high as 500 GHz. In order to improve the interaction between graphene and light, different strategies have been explored to enhance the built-in electric field formed near the metal electrode-graphene interfaces by applying electric field, chemical doping, metal nanoparticles, etc. For example, Kim et al. [61] demonstrated an all-graphene p–n vertical-type tunnelling photodiodes by p- and n-type doping of graphene, achieving a responsivity ($0.4 \sim 1.0$ A/W) and high detectivity ($\sim 10^{12}$ Jones) in the UV-NIR spectral range. Such a photoelectric performance comparable to those of commercial photodetectors implies a great potential for the applications as transparent and flexible optoelectronics.

However, the weak optical absorption of monolayer graphene (only $\sim 2.3\%$) and short lifetime of the photo-excited carriers still limit the responsivity of photodetectors. Naturally, the heterojunctions are expected to fabricate novel photodetectors by combining graphene with typical semiconductors, in which graphene plays main role in the separation and transport of photogenerated carrier and semiconductors mainly serve as optical absorbing materials. For example, the graphene/PbS quantum dot (QD) heterojunction photodetectors have been demonstrated to exhibit a responsivity of $\sim 10^7$ A/W and a specific detectivity of 7×10^{13} Jones by trapping the holes in the PbS QDs [62]. Remarkably, a highly sensitive MIR phototransistor based on graphene/plasmonic B-doped Si QD vdW heterojunctions has also been repotted, [63] in which the localized surface plasmon resonance of B-doped Si QDs enhanced the MIR absorption of graphene. The photogating effect induced by UV-NIR optical absorption of B-doped Si QDs results in the ultrabroadband detection, achieving an ultrahigh responsivity of $\sim 10^9$ A/W): and specific detectivity of $\sim 10^{13}$ Jones. Compared with 0D QDs, 2D layered semiconductors are commonly considered to be more suitable for the fabrication of graphene vdW heterojunction photodetectors due to larger atomic interface and stronger electron coupling effect. Nowadays, different transition metal dichalcogenides (TMDs): main group metal chalcogenides, and black phosphorus have been demonstrated to fabricate ultrathin graphene-based vdW heterojunction photodetectors in the few years [64–66]. For example, Li et al. [67] demonstrated a self-power graphene/h-BN/MoS_2 vdW heterojunction tunneling photodetector by inserting an atomically thin h-BN nanosheet into the graphene/MoS_2 interface. The interlayer carrier coupling under zero-bias can be effectively suppressed while the photogenerated holes can still transport via quantum tunneling, thereby achieving a high specific detectivity (5.9×10^{14} Jones for white-noise limited detectivity) and a high photoconversion efficiency (EQE > 80%). Moreover, different vdW configurations, such as graphene/2D semiconductor/graphene or N-type/graphene/P-type sandwich heterostructures, have also been demonstrated to significantly improve the photoelectric performance by enhancing the separation and charge transport in the vdW heterojunctions [68, 69].

Overall, SWCNTs and graphene have exhibited many novel exceptional photoelectronic properties and outstanding device performance, which are expected to play a key role in fabricating next-generation photoelectronic nanodevices. However, there are still many new challenges to overcome before possible practical application, including single-chirality SWCNTs, clean atomic interface, light-mater interaction, exciton dissociation, interface carrier dynamic process, vdW heterojunctions-induced electron coupling effect, and so on.

1.3.3 *Photovoltaic Devices*

As we know, SWCNTs can behave semiconducting or metallic and exhibit chirality-depended electrical and optical properties. Moreover, sc-SWCNTs are weak p-type direct-bandgap semiconductors in air with a bandgap inversely proportional to their diameter. Especially, high absorption coefficient ($10^4 \sim 10^5$ cm^{-1}) and wide light absorption within the solar spectrum make sc-SWCNT as an ideal light-harvesting material to fabricate high efficient photovoltaic devices, and an individual SWCNT photodiode with p–n junction has been demonstrated to exhibit remarkable photovoltaic effect and high power conversion efficiency (PCE) under illumination. For metallic SWCNTs and graphene, exceptional carrier mobility and high transparency also enable them great potential for novel photovoltaic devices as carrier collecting/transporting layer or transparent window.

Generally, SWCNTs are commonly used in the bulk photovoltaic devices as light harvesting materials, carrier transporter in active layer, charge separation and transporting layer, and so on. Previous studies usually focused on the incorporation of SWCNTs in the polymer bulk heterojunction photovoltaic cells, in which short-circuit current density (J_{SC}) and fill factor (FF) increased due to the enhanced charge separation efficiency, high carrier transportation ability and the reduction in the series resistance. Significantly, all-carbon photoactive layer composed of sc-SWCNT and electron-accepting fullerenes can also be used to fabricate novel carbon-based NIR photovoltaic cells, in which SWCNT/fullerene bulk or planar heterojunctions have usually been used as photoactive layers. Low-cost solution processing enable them serve as a coating on glass in buildings and windows to collect solar energy. However, it is worth noting that the existence of metallic SWCNTs or mixed chiral sc-SWCNTs influences the charge separation and transporting process in SWCNT films to some extent, because a high recombination rate usually happens due to the charge transfer between two SWCNTs induced by bandgap mismatch. In addition, SWCNTs as a hole transport layer have also been widely used to improve the photoelectric conversion process due to their p-type behavior and high carrier mobility. Especially, the photoelectric performance of perovskite solar cells has been improved significantly by introducing high quality SWCNTs [70–72].

Remarkably, SWCNTs have also been demonstrated to directly fabricate solar cells with traditional solar cell semiconductors (Si and GaAs): in which the distinct properties can be fully utilized by combining the high mobility and excellent photoelectric properties of SWCNTs with typical semiconductors, Since 2007, these novel heterojunction solar cells have attracted increasing attentions interest, which are easily fabricated by coating a semitransparent SWCNT film on n-type silicon to create p-n heterojunctions. For mixed or metallic SWCNTs, the nanotubes films as transparent conductive electrodes are mainly used to separate and transport the charged carriers at the interface due to the formation of a built-in potential, whereas sc-SWCNTs not only served as collecting/transport layer of charge carriers, but also contribute a certain photocurrent due to their light absorption. In the past decades, many efforts have been made to improve the PCE of SWCNT/Si heterojunction solar cells by optimizing the device structure, interface passivation, and chemical doping, and so on [73–75]. The PCE of heterojunction solar cells has rapidly increased to over 15%, and a new record PCEs of 18.9% has been demonstrated in SWCNT/Si solar cells with industry standard device geometries by a front and back-junction design, [76]. However, GaAs as a direct bandgap semiconductor is expected to fabricate a SWCNT-based heterojunction solar cell with a higher PCE due to high electron mobility and large optical absorption coefficient ($\sim 10^4$/cm). More importantly, the simple device configuration and easy fabrication enable the development of high-performance low-cost SWCNT/GaAs heterojunction solar cells solar cells. Although this type of solar cells has not been well investigated in the past, a high PCE of 10.15% has been demonstrated by coating (6, 5)-enriched sc-SWCNT films on the surface of n-type GaAs. It is believed that the SWCNT/GaAs vdW heterojunction solar cells exhibit higher PCE after optimizations due to similar working mechanism with SWCNT/Si heterojunctions.

Except for SWCNTs, the graphene/bulk semiconductor heterojunction solar cells have also been attracted increasing attentions, where high carrier mobility and high transparency makes graphene serve as both a carrier collecting/transporting layer and a transparent window. Similar with SWCNTs, the graphene/Si and graphene/GaAs heterojunction solar cells have also been developed by transferring highly transparent graphene film on the surface of n-type Si and GaAs. After optimizing the heterojunction interface, chemical doping of graphene and an anti-reflection layer, high efficiency graphene/Si and graphene/GaAs solar cells exhibited a PCE of 15.2 and 16.2% under standard illumination conditions, and it is expected to be further improve the PCE with the development of high-quality graphene and atomic clean interface technology.

1.3.4 Chemical Sensing Nanodevices

Large specific surface area and all surface atoms make SWCNTs sensitive to the surrounding environments, where the noncovalent interactions between absorbed molecules and SWCNTs result in a redistribution of charge carriers within SWCNTs, thereby changing the charge transport properties of SWCNTs. Nowadays, SWCNTs as the active element have been widely used to fabricate the chemical sensing nanodevices. The pioneering work about SWCNT gas sensors was made by Kong et al. in 2000, [19] in which the FET with an individual SWCNT as conducting channel was investigated when exposed to oxidative (NO_2) and reductive (NH_3) gases. The threshold voltages were observed to shift remarkably in the positive and negative directions for NO_2 and NH_3 exposure, respectively. High sensing performance demonstrated that SWCNT nanodevices have more promising potentials than metal oxide or polymer gas sensors. Since then, CNT-based chemical sensing nanodevices have attracted more and more research interests.

It has been demonstrated that the sensing performance is commonly dominated by several main factors, such as the surface states and conductive type of CNTs, molecule adsorption dynamics, charge transfer between molecules and CNTs, and Schottky barrier modulation at the CNT/metal contacts, etc. For example, the resistance changes in MWCNTs and mixed SWCNTs induced by molecule absorption are usually smaller than that in sc-SWCNTs, implying that sc-SWCNTs are more suitable for highly sensitive detections, whether for resistance type or FET type sensors [77–79]. And the surface states and curvature of sc-SWCNTs play a critical role in the adsorption–desorption dynamics of different molecules on the surfaces, further determining the gas sensing performance (sensitivity, selectivity and response time, etc.). Consequently, numerous efforts have been made to design and modify the surface of CNTs using specific chemical functional groups, achieving high sensitivity and selectivity toward various gases and vapors at room-temperature [80]. More importantly, the bandgap feature of sc-SWCNTs fundamentally determines the charge transfer dynamics between sc-SWCNTs and molecules, which has been used to theoretically guide the design and optimization of high-performance gas sensors. Nowadays, pure CNTs, functionalized CNTs and CNT-based heterojunctions as sensing materials have been extensively used to detect different kinds of gas and vapor molecules, and the detection sensitivity can be down to the single molecule level. After continuous efforts in the past two decades, CNT-based chemical sensors have been demonstrated to be a great application potential in different fields, such as pollution gas monitoring, rapid evaluation of food freshness and fruit ripeness, exhaled breath analysis, portable detection of chemical warfare agents and explosives, and so on [80]. With the development of the synthesis and separation of monochiral sc-SWCNTs, the chemical sensors based on specific sc-SWCNTs are expected to detect more gases or vapors with higher sensitivity and selectivity, which will be used in practice applications in the near future [81].

Except for SWCNTs, 2D graphene has also been extensively explored to detect gases and vapor sensors due to its large specific surface area. Schedin et al. [82] first demonstrated the thermally reversible detection of chemical gas/vapor molecules (NO_2, NH_3, CO, and H_2O) using bottom-gate graphene FETs, in which high sensitivity down to the single molecule level could be achieved by modulating the transverse Hall resistance at high magnetic field (10 T). Even though, the bandgap-free nature make graphene not suitable for high performance gas sensors based on the concentration-depended conductive changes. On the one hand, the conductivity change of graphene is very small before and after molecular adsorption and or charge transfer between them. On the other hand, the insufficient active sites on the surface for gas adsorption the interaction between gas molecules and graphene, thereby limiting the enhancement in sensitivity and selectivity of sensors.

Therefore, the rGO prepared by chemical method is gradually used as sensing materials for chemical sensors, because a large number of chemical functional groups exist on the surface of rGO sheets and the amount can be tailored by controlling the degree of reduction. Moreover, different functional groups can be flexibly designed and controllably modified on the surface of rGO by chemical functionalization, which shows stronger response to the target gas or vapor molecules with high sensitivity and selectivity. Similar to CNTs, different sensing materials based on rGO and its hybrids have been widely used to construct room-temperature chemical sensors, and these different-type sensors have also shown great potential for the applications in the abovementioned fields [37, 83].

1.3.5 Energy Storage Devices

Large surface area, ultrahigh charge mobility, superior conductivity, and high electrochemical activity make SWCNTs and graphene great potential in the novel electrochemical energy storage (EES) devices such as Li-ion batteries (LIBs) and supercapacitors, in which high energy density, high power density and long cycle life are highly required to satisfy the increasing demands in our energy-intensive society. Generally, SWCNTs, graphene and their associated derivatives have been widely utilized as either electrode materials or electrode additives for exploring high-performance LIBs and supercapacitors, because they can provide effective routes for the electron transfer and ion diffusion in EES devices. Nowadays, SWCNTs and rGO have been widely industrialized as conductive additives in the commercial LIBs, and it is predicated that the demand for only SWCNT raw materials will exceed 250 tons per year in 2025. With the development of low-cost mass production techniques, more carbon nanomaterials will be commercially used in the EES devices to achieve higher electrochemical performance.

For 1D SWCNTs, the theoretical specific capacity can exceed 1116 mA h g^{-1} and the Li ions diffuse into the stable sites located on the surface, inside individual nanotubes and between SWCNTs [84, 85]. However, the electrochemical response of SWCNTs towards Li ions is strongly dependent on the surface state of nanotubes, and the practical capacity is usually s less than 200 mAh g^{-1} after 100 cycles owing to large irreversible capacity and high voltage hysteresis [86]. Therefore, chemical treatments have been extensively used to introduce a large number of defects on the wall of nanotubes, providing additional sites for Li ion adsorption and then improving the practical capacity of SWCNTs [87, 88]. In practice applications, well-dispersed SWCNTs as conductive additives are directly mixed with the active materials by simple mixing process, and ultrahigh aspect ratio of SWCNTs enable the formation of a conductive network at a low weight loading. Meanwhile, SWCNT-based composites have also been widely reported as novel active materials for high performance LIBs, in which the functionalized SWCNTs can be used as a template to form nanocomposites with the common-used cathode materials, such as $LiCoO_2$, $LiFePO_4$, spinel $LiMn_2O_4$, and $LiNi_xCo_yMn_{1-x-y}O_2$.

Consequently, the SWCNTs on the surface of cathode particles form a 3D conductive network, thereby enabling the fast charge transfer capability and high energy density with long durability [89–91]. Meanwhile, similar improvement strategy have also been extensively utilized by combining SWCNTs with popular used anode materials, which exhibit a higher theoretical specific capacity but worse Li-ions cyclic stability than commercial graphite anode. Especially, the low conductivity and large volume change of Si anode materials during charge/discharge process have seriously limited the commercial application in LIBs. By introducing SWCNTs into the Si anode, both the abovementioned critical issues can be well alleviated, [92–94] which have also been demonstrated for transition metal oxides or sulfides as anode materials [95–97]. In addition, SWCNTs are also able to be assembled into free-standing electrodes (without any binder or current collector) as a physical support or current collectors for ultra-high capacity electrode materials like silicon. SWCNTs network can release the strain of a large volume change during lithium insertion/extraction and result in anode with an excellent cyclability. Compared to the anode of active material on Cu current collector, free-standing SWCNTs based electrodes can decrease the weight percentage of the anode side and significantly improve the energy density of LIBs [98].

Similarly, the applications of 2D graphene and its derivatives in LIBs have also attracted continuous attentions in the past decades. Although the massive irreversibility of the first lithiation step, layer restacking and high voltage hysteresis are also the crucial issues that affect the applications of graphene as active materials for LIBs, numerous efforts have been made to improve the EES behavior and performance, including pre-lithiation, controlled surface functionalization and the use of nanocomposites [99]. In particular, rGO has been demonstrated to exhibit superior electrochemical capacity surpassing the theoretical value of single-layer graphene due to the existence of oxygenated groups and defects, and numerous rGO-based naoncomposites achieve enhanced electrochemical capabilities by incorporating metal oxide/sulfide on the surface of rGO nanosheets to prevent restacking of the graphene sheets. In fact, graphene-based nanocomposites have shown outstanding performance in LIBs. Graphene promises to increase the energy and power density of practical systems, as well as enable the development of next-generation LIBs. However, the results so far suggest that real breakthroughs are still to come.

Different from LIBs, the electrochemical supercapacitors (SCs) can store and release the energy at relatively high rates with high charge/discharge efficiency close to 100%, demonstrating outstanding power performance, excellent reversibility, ultra-long cycle life (>10^6 cycles): and simple operation, and so on. Although the specific energy of SCs is somewhat low, the specific power is higher than that of most LIBs. Moreover, both the specific energy and specific power of SCs can be controlled within several orders of magnitude by changing the device designs, making them to meet the increasing power demands as a stand-alone energy supply or a hybrid system combining with LIBs. For SWCNTs and graphene, high specific surface area and charge transport capability make them to be promising electrode materials for high performance SCs [2, 100–102]. However, carbon-based composites are usually adopted to improve the energy and power densities of SCs [103, 104]. For example, the mixture of activated carbons and SWCNTs has been commonly used to enhance the electrical conductivity and mechanical strength of electrodes. In addition, SWCNTs have also been demonstrated to remarkably improve the charge transfer rate and the charge storage capability by combining with conducting polymer and/or transition metal oxides, such as polyaniline (PANI): polypyrrole (PPy): poly (3,4-ethylenedioxythiophene) (PEDOT): MnO_2, In_2O_3, and so on [105–108]. Similarly, in order to achieve a high-performance graphene-based SCs, graphene or rGO-based composites have also commonly synthesized by combining the pseudocapacitive nanomaterials such as metal oxides and sulfides. In the composite system, fast charge transfer can be obtained through flexible conducting networks of graphene on the surface of pseudocapacitive materials, while these oxides or sulfides also can effectively prevent the aggregation of graphene sheets. By taking advantage of their synergy effect of the different components in the active materials, the EES performances of SCs are expected to be further improved with the development of carbon materials and composite technologies [23, 109].

1.3.6 Thermal Management

With the development of 5G, cloud computing technologies and the electronic integration in electronics industry, highly efficient heat dissipation has become a crucial issue. Meanwhile, high thermal conductivity materials are also required to spread the heat generated locally in the optoelectronic devices, where the reliability and speed are strongly affected by temperature. However, the traditional heat dissipation materials (Cu or Al) cannot satisfy the technical requirements. As we know, SWCNTs and graphene possess exceptional thermal properties due to the strong covalent bonds between carbon atoms, the thermal conductivity of individual SWCNTs is as high as $\sim 3000\ \mathrm{W\ m^{-1}\ K^{-1}}$ and suspended graphene has the highest thermal conductivity of $5000\ \mathrm{W\ m^{-1}\ K^{-1}}$ at room temperature among natural materials. Therefore, SWCNTs and graphene are promising materials for thermal management of electronic equipment [110–112]. Especially, free-standing graphene film has been industrialized to enhance the heat dissipation in mobile phones and other electronic equipment. However, the anisotropic thermal conductivity results in a low heat dissipation capability in the direction perpendicular to graphene film, which also exists in the SWCNT film. As a result, increasing the longitudinal thermal conductivity of film is very important for practical application. Although vertical MWCNT arrays as thermal interface materials have been demonstrated to promote the interfacial thermal transport, [113] vertical SWCNT arrays are rarely used due to high cost and complex preparation technology. SWCNTs are most widely used to improve the thermal conductivity of thermal conductive polymer by optimizing the loading amount. However, with the development of preparation technology for carbon-based heat dissipation materials and structures, we believe that SWCNTs and graphene are expected to play a dominant role in solving the problem of heat dissipation in the near future.

1.3.7 Biology

Ultrahigh specific surface area, exceptional physical and chemical properties also make SWCNTs and graphene have great potential in the biological and medical applications. For example, SWCNTs as nanocarriers have been demonstrated to selectively deliver plasmid DNA to chloroplasts of different plant species deliver or small interfering RNA (siRNA) into the cells of plant tissues due to small diameter, high-aspect ratio, and high stiffness [114, 115]. The sc-SWCNTs are extensively utilized as multifunctional nano-probes for biomedical imaging or cancer therapy due to large scattering cross-sections and suitable fluorescence wavelength within the NIR II widow, and unique thermal properties. More importantly, the electronic or optical properties are very sensitive to the change in the local environment on the surface of SWCNTs, enabling the novel biomolecule detection after chemical modification and bio-functionalization. Compared with MWCNTs, SWCNTs with all atoms on the surface are more suitable to be used in the different type biosensors due to the chiral-related electronic and optical properties. By choosing specific sc-SWCNTs and functionalization modifications, a large number SWCNT-based biosensors have been explored to detect different molecular analytes with and high sensitivity and selectivity, such as DNA sequences, proteins, enzyme, antigen/ antibody molecules, biomarkers of diseases, and so on [116, 117]. Nowadays, SWCNT-based biosensors are usually be divided into chemical biosensors and physical biosensors according to the interactions between modified SWCNTs and analytes. Among them, the SWCNT-based FETs and fluorescence-related optical biosensors have attracted increasing attentions. In particular, with the development of chirality-selective synthesis of sc-SWCNTs, the monochiral nanotubes with high fluorescence efficiency are expected to fabricate next-generation high-performance SWCNT-based FETs and optical biosensors, because the carrier recombination probability induced by charge transfer between nanotubes is reduced.

Similar to SWCNTs, graphene-based biosensors have also been demonstrated by using the same functionalization strategies. Especially, GO and rGO are extensively used for biomolecule recognition, bioassays, molecular medicine and small molecular drug delivery due to large surface area, easy functionalization, high conductivity and good biocompatibility [118]. Until now, many research works highlight the applications of graphene materials for tissue engineering, molecular drug delivery, cancer treatment, biosensing, and so on [119, 120]. It is worth noting that, although some results suggest that biomolecule coating modification can reduce toxicity, the health and environmental risks are still the major concerns before practical applications [121].

1.4 Future Perspective of Carbon Nanotubes and Graphene

Until now, only MWCNTs and rGO have been commercially used in large scale in the industry, such as conductive additive in Li-ion battery and antistatic polymer, mechanical enhancers in rubber, heat dissipating material for mobile phones, as so on. However, a few SWCNT production has been demonstrated to exhibit higher performance and balanced cost performance, which are expected to gradually replace MWCNTs and rGO in some high-end applications. In particular, with the further development of mass production technology of SWCNTs, we believe that SWCNTs will be more widely applied in different commercial fields, resulting in an exceptional property enhancement or endowing a new product performance. Especially, energy storage and anti-static electricity fields are expected to have greater development opportunities in the next few years.

More importantly, a series of major breakthroughs have been made in the chirality-selective synthesis of sc-SWCNTs, promoting significantly the development and applications of SWCNTs in different fields, such as electronics, optoelectronics, sensors, biology, etc. Especially, some important advances in SWCNT-based nanoelectronics and optoelectronics are expected to be made based on the preparation technology of wafer-scale sc-SWCNT film in the next few years, and high performance SWCNT-based sensors will also be explored for potential applications. However, it is worth noting that the basic research on growth and process of sc-SWCNTs still needs to be strengthened, especially on the growth mechanism and mass production of monochiral SWCNTs, high-density sc-SWCNT arrays, wafer-scale efficient assembly, and efficient separation of large-diameter monochiral SWCNT, and so on. For example, monochiral sc-SWCNTs with a purity of >99.999% is required to meet the demand of large-scale nanoelectronics. After that, SWCNT-based nanodevices and nanosystems will be used in practical applications.

For graphene, the progress of wafer-size growth techniques makes graphene to replace ITO as the novel transparent conductive film in the display field. More importantly, high-quality graphene has been demonstrated to fabricated high performance high-frequency electronics and optoelectronics due to ultrahigh carrier mobility and wide band photoelectric response. It is believed that graphene-based RF and optoelectronic nanodevices would be widely used in different fields, including the next-generation RF, terahertz, optical communication, infrared imaging, and so on. In addition, combining with the exceptional physical and chemical properties of sc-SWCNTs, these carbon nanomaterials enable us to fabricate novel and flexible all-carbon electronic, optoelectronic, and sensing nanodevices, which are expected to exhibit a great potential for next-generation transparent nanodevices and nanosystems.

References

1. Jariwala D, Sangwan VK, Lauhon LJ, Marks TJ, Hersam MC (2013). Carbon nanomaterials for electronics, optoelectronics, photovoltaics, and sensing. Chem Soc Rev, 42: 2824–2860.
2. Dai LM, Chang DW, Baek JB, Lu W (2012). Carbon nanomaterials for advanced energy conversion and storage. Small, 8(8):1130–1166.
3. Hirsch A (2010). The era of carbon allotropes. Nat Mater, 9(11): 868–871.
4. Ding L, Zhang Z, Liang S, Pei T, Wang S, Li Y, Zhou W, Liu J, Peng L (2012). CMOS-based carbon nanotube pass-transistor logic integrated circuits. Nat Commun, 3: 677.
5. Cao Q (2021). Carbon nanotube transistor technology for More-Moore scaling. Nano Res, 1–19.
6. Du X, Skachko I, Barker A, Andrei EY (2008). Approaching ballistic transport in suspended graphene. Nat Nanotech, 3(8): 491–495.
7. Perebeinos V, Avouris P (2008). Phonon and electronic nonradiative decay mechanisms of excitons in carbon nanotubes. Phys Rev Lett, 101(5): 057401.
8. Wen C, Li J, Kitazawa K, Aida T, Honma I, Komiyama H, Yamada K (1992). Electrical conductivity of a pure C_{60} single crystal. Appl Phys Lett, 61(18): 2162–2163.
9. Krätschmer W, Lamb LD, Fostiropoulos K, Huffman DR (1990). Solid C $_{60}$: a new form of carbon. Nature, 347(6291): 354–358.
10. Zhu SE, Li F, Wang GW (2013). Mechanochemistry of fullerenes and related materials. Chem Soc Rev, 42(18): 7535–7570.
11. Georgakilas V, Perman JA, Tucek J, Zboril R. (2015). Broad family of carbon nanoallotropes: classification, chemistry, and applications of fullerenes, carbon dots, nanotubes, graphene, nanodiamonds, and combined superstructures. Chem Rev, 115(11): 4744–4822.
12. Castro E, Garcia AH, Zavala G, Echegoyen L (2017). Fullerenes in biology and medicine. J Mater Chem B, 5(32): 6523–6535.
13. Jun GH, Jin SH, Lee B, Kim BH, Chae WS, Hong SH, Jeon S (2013). Enhanced conduction and charge-selectivity by N-doped graphene flakes in the active layer of bulk-heterojunction organic solar cells. Energy Environ Sci, 6(10): 3000–3006.
14. Cui Y, Yao H, Zhang J, Xian K, Zhang T, Hong L, Wang Y, Xu Y, Ma K, An C, He C, Wei Z, Gao F, Hou J (2020). Single-junction organic photovoltaic cells with approaching 18% efficiency. Adv Mater, 32(19): 1908205.
15. Xu X, Ray R, Gu Y, Ploehn HJ, Gearheart K, Raker K, Scrivens WA (2004). Electrophoretic analysis and purification of fluorescent single-walled carbon nanotube fragments, J Am Chem Soc, 126: 12736–12737.
16. Takakura A, Beppu K, Nishihara T, Fukui A, Kozeki T, Namazu T, Miyauchi Y, Itami K. (2019). Strength of carbon nanotubes depends on their chemical structures. Nature Commun, 10(1): 1–7.
17. Cai BF, Su YJ, Tao ZJ, Hu J, Zou C, Yang Z, Zhang YF (2018). Highly sensitive broadband single-walled carbon nanotube photodetectors en-hanced by separated graphene nanosheets. Adv Optical Mater, 6(23): 1800791.
18. Avouris P, Freitag M, Perebeinos V (2008). Carbon-nanotube photonics and optoelectronics. Nat Photonics, 2(6): 341–350.
19. Kong J, Franklin NR, Zhou C, Chapline MG, Peng S, Cho K, Dai HJ (2000). Nanotube molecular wires as chemical sensors. Science, 287(5453): 622–625.
20. Gaviria Rojas WA, Hersam MC (2020). Chirality-enriched carbon nanotubes for next-generation computing. Adv. Mater., 32(41): 1905654.
21. Liu Z, Tabakman S, Welsher K, Dai H (2009). Carbon nanotubes in biology and medicine: in vitro and in vivo detection, imaging and drug delivery. Nano Res, 2(2): 85–120.
22. Hecht DS, Hu L, Irvin G (2011). Emerging transparent electrodes based on thin films of carbon nanotubes, graphene, and metallic nanostructures. Adv Mater, 23(13): 1482–1513.

23. Ni J, Li Y (2016). Carbon nanomaterials in different dimensions for electrochemical energy storage. Adv Energy Mater, 6(17): 1600278.
24. Xu ZL, Kim JK, Kang K (2018). Carbon nanomaterials for advanced lithium sulfur batteries. Nano Today, 19: 84–107.
25. Novoselov KS, Geim AK, Morozov SV, Jiang D, Zhang Y, Dubonos SV, Grigorieva IV, Firsov AA (2004). Electric field effect in atomically thin carbon films. Science, 306(5696): 666–669.
26. Nika DL, Pokatilov EP, Askerov AS, Balandin AA (2009). Phonon thermal conduction in graphene: Role of Umklapp and edge roughness scattering. Phys Rev B, 79(15): 155413.
27. Zhang J, Lin L, Jia K, Sun L, Peng H, Liu Z (2020). Controlled growth of single-crystal graphene films. Adv Mater, 32(1): 1903266.
28. Yi M, Shen Z (2015). A review on mechanical exfoliation for the scalable production of graphene. J Mater Chem A, 3(22): 11700–11715.
29. Chen J, Yao B, Li C, Shi G (2013). An improved Hummers method for eco-friendly synthesis of graphene oxide. Carbon, 64: 225–229.
30. Yang W, Chen G, Shi Z, Liu CC, Zhang L, Xie G, Cheng M, Wang D, Yang R, Shi D, Watanabe K, Taniguchi T, Yao Y, Zhang Y, Zhang G (2013). Epitaxial growth of single-domain graphene on hexagonal boron nitride. Nat Mater, 12(9): 792–797.
31. Li X, Cai W, An J, Kim SY, Nah J, Yang DX, Piner R, Velamakanni A, Jung I, Tutuc E (2009). Large-area synthesis of high-quality and uniform graphene films on copper foils. Science, 324(5932): 1312–1314.
32. Hernandez Y, Nicolosi V, Lotya M, Blighe FM, Sun Z, De S, McGovern IT, Holland B, Byrne M, Gun'Ko YK, Boland JJ, Niraj P, Duesberg G, Krishnamurthy S, Goodhue R, Hutchison J, Scardaci V, Ferrari AC, Coleman JN (2008). High-yield production of graphene by liquid-phase exfoliation of graphite. Nature Nanotechnol, 3(9): 563–568.
33. Marcano DC, Kosynkin DV, Berlin JM, Sinitskii A, Sun Z, Slesarev A, Sun ZZ, Slesarev A, Alemany LB, Lu W, Tour JM, Tour JM (2010). Improved synthesis of graphene oxide. ACS Nano, 4(8): 4806–4814.
34. Compton OC, Nguyen ST (2010). Graphene oxide, highly reduced graphene oxide, and graphene: versatile building blocks for carbon-based materials. Small, 6(6): 711–723.
35. Zhu Y, Murali S, Cai W, Li X, Suk JW, Potts JR, Ruoff RS (2010). Graphene and graphene oxide: synthesis, properties, and applications. Adv Mater, 22(35): 3906–3924.
36. Wang B, Ruan T, Chen Y, Jin F, Peng L, Zhou Y, Wang D, Dou S (2020). Graphene-based composites for electrochemical energy storage. Energy Storage Mater, 24: 22–51.
37. Singh E, Meyyappan M, Nalwa HS (2017). Flexible graphene-based wearable gas and chemical sensors. ACS Appl Mater Interfaces, 9(40): 34544–34586.
38. Li G, Li Y, Liu H, Guo Y, Li Y, Zhu D (2010). Architecture of graphdiyne nanoscale films. Chem Commun, 46: 3256–3258.
39. Li Y, Xu L, Liu H, Li Y (2014). Graphdiyne and graphyne: from theoretical predictions to practical construction. Chem Soc Rev, 43: 2572–2586.
40. Ge C, Chen J, Tang S, Du Y, Tang N (2019). Review of the electronic, optical, and magnetic properties of graphdiyne: from theories to experiments. ACS Appl Mater Interfaces, 11(3): 2707–2716.
41. Peng LM, Zhang Z, Wang S (2014). Carbon nanotube electronics: recent advances. Mater Today, 17(9): 433–442.
42. Hills G, Lau C, Wright A, Fuller S, Bishop MD, Srimani T, Kanhaiya P, Ho R, Amer A, Stein Y, Murphy D, Arvind, Chandrakasan A, Shulaker MM (2019). Modern microprocessor built from complementary carbon nanotube transistors. Nature, 572(7771): 595–602.
43. Liu L, Han J, Xu L, Zhou J, Zhao C, Ding S, Shi H, Xiao M, Ding L, Ma Z, Jin C, Zhang Z, Peng L (2020). Aligned, high-density semiconducting carbon nanotube arrays for high-performance electronics. Science, 368(6493): 850–856.
44. Shi H, Ding L, Zhong D, Han J, Liu L, Xu L, Sun P, Wang H, Zhou J, Fang L, Zhang Z, Peng L (2021). Radiofrequency transistors based on aligned carbon nanotube arrays. Nature Electronics, 4(6): 405–415.

45. Kong W, Kum H, Bae SH, Shim J, Kim H, Kong L, Meng Y, Wang K, Kim C, Kim J (2019). Path towards graphene commercialization from lab to market. Nature Nanotech, 14 (10): 927–938.
46. Wu Y, Lin Y, Bol AA, Jenkins KA, Xia F, Farmer DB, Zhu Y, Avouris P (2011). High-frequency, scaled graphene transistors on diamond-like carbon. Nature, 472: 74–78.
47. Cheng R, Bai JW, Liao L, Zhou HL, Chen Y, Liu LX, Lin YC, Jiang S, Huang Y, Duan XF (2012). High-frequency self-aligned graphene transistors with transferred gate stacks. Proc Nat Acad Sci, 109: 11588–11592.
48. Wu Y, Zou X, Sun M, Cao Z, Wang X, Huo S, Zhou JJ, Yang Y, Yu XX, Kong YC, Yu GH, Liao L, Chen T (2016). 200 GHz Maximum oscillation frequency in CVD graphene radio frequency transistors. ACS Appl Mater Interfaces, 8: 25645–25649.
49. He X, Léonard F, Kono J (2015). Uncooled carbon nanotube photodetectors. Adv Optical Mater, 3(8): 989–1011.
50. Ahn YH, Tsen AW, Kim B, Park YW, Park J (2007). Photocurrent imaging of p-n junctions in ambipolar carbon nanotube transistors. Nano Lett, 7(11): 3320–3323.
51. Yang L, Wang S, Zeng Q, Zhang Z, Peng LM (2013). Carbon nanotube photoelectronic and photovoltaic devices and their applications in infrared detection. Small, 9(8): 1225–1236.
52. Liu Y, Wei N, Zeng Q, Han J, Huang H, Zhong D, Wang F, Ding L, Xia J, Xu H, Ma Z, Qiu S, Li Q, Liang X, Zhang Z, Wang S, Peng L (2016). Room temperature broadband infrared carbon nanotube photodetector with high detectivity and stability. Adv Optical Mater, 4(2): 238–245.
53. Itkis ME, Borondics F, Yu A, Haddon RC (2006). Bolometric infrared photoresponse of suspended single-walled carbon nanotube films. Sci-ence, 312(5772): 413–416.
54. St-Antoine BC, Ménard D, Martel R (2009). Position sensitive pho-tothermoelectric effect in suspended single-walled carbon nanotube films. Nano Lett, 9(10): 3503–3508.
55. Huo T, Yin H, Zhou D, Sun L, Tian T, Wei H, Hu N, Yang, Zhang Y, Su Y (2020). Self-powered broadband photodetector based on single-walled carbon nanotube/GaAs heterojunctions. ACS Sustainable Chem Eng, 8(41): 15532–15539.
56. Flemban TH, Haque MA, Ajia I, Alwadai N, Mitra S, Wu T, Roqan IS (2017). A photodetector based on p-Si/n-ZnO nanotube heterojunctions with high ultraviolet responsivity. ACS Appl Mater Interfaces, 9(42): 37120–37127.
57. Riaz A, Alam A, Selvasundaram PB, Dehm S, Hennrich F, Kappes MM, Krupke R (2019). Near-infrared photoresponse of waveguide-integrated carbon nanotube–silicon junctions Adv Electron Mater, 5(1): 1800265.
58. Park S, Kim SJ, Nam JH, Pitner G, Lee TH, Ayzner AL, Wang HL, Fong SW, Vosgueritchian M, Park YJ, Brongersma ML, Bao ZN (2015). Sig-nificant enhancement of infrared photodetector sensitivity using a semi-conducting single-walled carbon nanotube/ C_{60} phototransistor. Adv Mater, 27: 759–765.
59. Bergemann K, Léonard F (2018). Room-temperature phototransistor with negative photoresponsivity of 10^8 AW^{-1} using fullerene-sensitized aligned carbon nanotubes. Small, 14(42): 1802806.
60. Liu Y, Wang F, Wang X, Wang X, Flahaut E, Liu X, Li Y, Wang X, Xu Y, Shi Y, Zhang R (2015). Planar carbon nanotube–graphene hybrid films for high-performance broadband photodetectors. Nat Commun, 6: 8589.
61. Kim CO, Kim S, Shin DH, Kang SS, Kim JM, Jang CW, Joo SS, Lee JS, Kim JH, Choi SH, Hwang E (2014). High photoresponsivity in an all-graphene p–n vertical junction photodetector. Nature Commun, 5(1): 1–7.
62. Konstantatos G, Badioli M, Gaudreau L, Osmond J, Bernechea M, de Arquer FPG, Gatti F, Koppens FHL (2012). Hybrid graphene–quantum dot phototransistors with ultrahigh gain. Nat Nanotechol, 7: 363–368.
63. Ni Z, Ma L, Du S, Xu Y, Yuan M, Fang H, Wang Z, Xu M, Li D, Yang J, Hu W, Pi X, Yang D (2017). Plasmonic silicon quantum dots enabled high-sensitivity ultrabroadband photodetection of graphene-based hybrid phototransistors. ACS Nano, 11(10): 9854–9862.

64. Zhang W, Chuu CP, Huang JK, Chen CH, Tsai ML, Chang YH, Liang CT, Chen YZ, Chueh YL, He JH, Chou MY, Li LJ (2014). Ultrahigh-gain photodetectors based on atomically thin graphene-MoS_2 heterostructures. Sci Rep, 4: 3826.
65. Luo W, Cao Y, Hu P, Cai K, Feng Q, Yan F, Yan T, Zhang X, Wang K (2015). Gate tuning of high-performance InSe-based photodetectors using graphene electrodes. Adv Optical Mater, 3: 1418–1423.
66. Liu Y, Shivananju BN, Wang Y, Zhang Y, Yu W, Xiao S, Sun T, Ma W, Mu H, Lin S, Zhang H (2017). Highly efficient and air stable infrared photodetector based on 2D layered graphene-black phosphorus heterostructure. ACS Appl Mater Interfaces, 9(41): 36137–36145.
67. Li H, Li X, Park JH, Tao L, Kim KK, Lee YH, Xu JB (2019). Restoring the photovoltaic effect in graphene-based van der Waals heterojunctions towards self-powered high-detectivity photodetectors. Nano Energy, 57: 214–221.
68. Yu WJ, Liu Y, Zhou H, Yin A, Li Z, Huang Y, Duan XF (2013). Highly efficient gate-tunable photocurrent generation in vertical heterostructures of layered materials. Nat Nanotechnol, 8(12): 952–958.
69. Li A, Chen Q, Wang P, Gan Y, Qi T, Wang P, Tang F, Wu JZ, Chen R, Zhang L, Gong Y (2019). Ultrahigh-sensitive broadband photodetectors based on dielectric shielded MoTe/graphene/SnS p-g-n junctions. Adv Mater, 31(6): 1805656–1805664.
70. Luo Q, Wu RG, Ma LT, Wang CH, Liu H, Lin H, Wang N, Chen Y, Guo ZH (2021). Recent advances in carbon nanotube utilizations in perovskite solar cells. Adv Funct Mater, 31(6): 2004765.
71. Singh R, Singh P K, Bhattacharya B, Rhee HW (2019). Review of current progress in inorganic hole-transport materials for perovskite solar cells. Appl Mater Today, 14: 175–200.
72. Wang Y, Zhao H, Mei Y, Liu H, Wang S, Li X (2018). Carbon nanotube bridging method for hole transport layer-free paintable carbon-based perovskite solar cells. ACS Appl Mater Interfaces, 11(1): 916–923.
73. Tune DD, Flavel BS (2018). Advances in carbon nanotube–silicon hetero-junction solar cells. Adv Energy Mater, 8(15): 1703241.
74. Cui K, Anisimov AS, Chiba T, Fujii S, Kataura H, Nasibulin AG, Chiashi S, Kauppinen EI, Maruyama S (2014). Air-stable high-efficiency solar cells with dry-transferred single-walled carbon nanotube films. J Mater Chem A, 2: 11311–11318.
75. Tune DD, Mallik N, Fornasier H, Flavel BS (2020). Breakthrough carbon nanotube-silicon heterojunction solar cells. Adv Energy Mater, 10(1): 1903261.
76. Chen J, Tune DD, Ge K, Li H, Flavel BS (2020). Front and back-junction carbon nanotube-silicon solar cells with an industrial architecture. Adv Funct Mater, 30(17): 2000484.
77. Meyyappan M (2016). Carbon nanotube-based chemical sensors. Small, 12(16): 2118–2129.
78. Sacco L, Forel S, Florea I, Cojocaru CS (2020). Ultra-sensitive NO_2 gas sensors based on single-wall carbon nanotube field effect transistors: Monitoring from ppm to ppb level. Carbon, 157: 631–639.
79. Zhou S, Xiao M, Liu F, He J, Lin Y, Zhang Z (2021). Sub-10 parts per billion detection of hydrogen with floating gate transistors built on semiconducting carbon nanotube film. Carbon, 180: 41–47.
80. Schroeder V, Savagatrup S, He M, Lin S, Swager TM (2018). Carbon nanotube chemical sensors. Chem Rev, 119(1): 599–663.
81. Forel S, Sacco L, Castan A, Florea I, Cojocaru CS (2021). Simple and rapid gas sensing using a single-walled carbon nanotube field-effect transistor-based logic inverter. Nanoscale Adv, 2021, 3, 1582–1587.
82. Schedin F, Geim AK, Morozov SV, Hill EW, Blake P, Katsnelson MI, Novoselov KS (2007). Detection of individual gas molecules adsorbed on graphene. Nat Mater. 6(9): 652–655.

83. Liu Y, Dong X, Chen P (2012). Biological and chemical sensors based on graphene materials. Chem Soc Rev, 41(6): 2283–2307.
84. Zhao J, Buldum A, Han J, Lu JP (2000). First-principles study of Li-intercalated carbon nanotube ropes. Phys Rev Lett, 85(8): 1706.
85. Meunier V, Kephart J, Roland C, Bernholc J (2002). Ab initio investigations of lithium diffusion in carbon nanotube systems. Phys Rev Lett, 88(7): 075506.
86. Noerochim L, Wang JZ, Chou SL, Wexler D, Liu HK (2012). Free-standing single-walled carbon nanotube/SnO_2 anode paper for flexible lithium-ion batteries. Carbon, 50(3): 1289–1297.
87. Bulusheva LG, Okotrub AV, Kurenya AG, Zhang H, Zhang H, Chen X, Song H (2011). Electrochemical properties of nitrogen-doped carbon nanotube anode in Li-ion batteries. Carbon, 49(12): 4013–4023.
88. Wu YP, Rahm E, Holze R (2003). Carbon anode materials for lithium ion batteries. J Power Sources, 114(2): 228–236.
89. Kucinskis G, Bajars G, Kleperis J (2013). Graphene in lithium ion battery cathode materials: A review. J Power Sources, 240: 66–79.
90. Tang J, Zhong X, Li H, Li Y, Pan F, Xu B (2019). In-situ and selectively laser reduced graphene oxide sheets as excellent conductive additive for high rate capability $LiFePO_4$ lithium ion batteries. J Power Sources, 412: 677–682.
91. Wang GP, Zhang QT, Yu ZL, Qu MZ (2008). The effect of different kinds of nano-carbon conductive additives in LIB on the resistance and electrochemical behavior of the $LiCoO_2$ composite cathodes. Solid State Ionics, 179(7–8): 263–268.
92. Xiao Q, Fan Y, Wang X, Susantyoko RA, Zhang Q (2014). A multilayer Si/CNT coaxial nanofiber LIB anode with a high areal capacity. Energy Environ Sci, 7(2): 655–661.
93. Wang S, Liao J, Wu M, Xu Z, Gong F, Chen C, Wang Y, Yan X (2017). High rate and long cycle life of a CNT/rGO/Si nanoparticle composite anode for lithium-ion batteries. Particle Particle Syst Characteriz, 34(10): 1700141.
94. Cai H, Han K, Jiang H, Wang J, Liu H (2017). Self-standing silicon-carbon nanotube/ graphene by a scalable in situ approach from low-cost Al-Si alloy powder for lithium ion batteries. J Phys Chem Solid, 109: 9–17.
95. Lee WJ, Maiti UN, Lee JM, Lim J, Han TH, Kim SO (2014). Nitrogen-doped carbon nanotubes and graphene composite structures for energy and catalytic applications. Chem Commun, 50(52): 6818–6830.
96. Meng X, Riha SC, Libera JA, Wu Q, Wang HH, Martinson AB, Elam JW (2015). Tunable core-shell single-walled carbon nanotube-Cu_2S networked nanocomposites as high-performance cathodes for lithium-ion batteries. J Power Sources, 280: 621–629.
97. Tamate R, Saruwatari A, Nakanishi A, Matsumae Y, Ueno K, Dokko K, Watanabe M (2019). Excellent dispersibility of single-walled carbon nanotubes in highly concentrated electrolytes and application to gel electrode for Li-S batteries. Electrochem Commun, 109: 106598.
98. Cui LF, Hu L, Choi JW, Cui Y (2010). Light-weight free-standing carbon nanotube-silicon films for anodes of lithium ion batteries. ACS Nano, 4(7): 3671–3678.
99. Caballero Á, Morales J (2012). Can the performance of graphene nanosheets for lithium storage in Li-ion batteries be predicted? Nanoscale, 4(6): 2083–2092.
100. Kaempgen M, Chan C K, Ma J, Cui Y, Gruner G (2009). Printable thin film supercapacitors using single-walled carbon nanotubes. Nano Lett, 9(5): 1872–1876.
101. Niu Z, Dong H, Zhu B, Li J, Hng HH, Zhou WY, Chen XD, Xie SS (2013). Highly stretchable, integrated supercapacitors based on single-walled carbon nanotube films with continuous reticulate architecture. Adv Mater, 25(7): 1058–1064.
102. Niu Z, Zhou W, Chen J, Feng G, Li H, Ma WJ, Li JZ, Dong HB, Ren Y, Zhao D, Xie SS (2011). Compact-designed supercapacitors using free-standing single-walled carbon nanotube films. Energy Environ Sci, 4(4): 1440–1446.
103. Cheng Q, Tang J, Ma J, Zhang H, Shinya N, Qin LC (2011). Graphene and carbon nanotube composite electrodes for supercapacitors with ultra-high energy density. Phys Chem Chem Phys, 13(39): 17615–17624.

104. Bose S, Kuila T, Mishra AK, Rajasekar R, Kim NH, Lee JH (2012). Carbon-based nanostructured materials and their composites as supercapacitor electrodes. J Mater Chem, 2012, 22(3): 767–784.
105. Gupta V, Miura N (2006). Polyaniline/single-wall carbon nanotube (PANI/SWCNT) composites for high performance supercapacitors. Electrochimica Acta, 52(4): 1721–1726.
106. Chen PC, Shen GZ, Shi Y, Chen HT, Zhou CW (2010). Preparation and characterization of flexible asymmetric supercapacitors based on transition-metal-oxide nanowire/single-walled carbon nanotube hybrid thin-film electrodes. ACS Nano, 4(8): 4403–4411.
107. Zhi M, Xiang CC, Li JT, Li M, Wu NQ (2013). Nanostructured carbon–metal oxide composite electrodes for supercapacitors: a review. Nanoscale, 5(1): 72–88.
108. Antiohos D, Folkes G, Sherrell P, Ashraf S, Wallace GG, Aitchison P, Harris AT, Chen J, Minett AI (2011). Compositional effects of PEDOT-PSS/single walled carbon nanotube films on supercapacitor device performance. J Mater Chem, 21(40): 15987–15994.
109. Yang Z, Tian J, Yin Z, Cui C, Qian W, Wei F (2019). Carbon nanotube-and graphene-based nanomaterials and applications in high-voltage supercapacitor: a review. Carbon, 141: 467–480.
110. Xu J, Fisher TS (2006). Enhancement of thermal interface materials with carbon nanotube arrays. Inter J Heat Mass Transfer, 49(9–10): 1658–1666.
111. Shaikh S, Li L, Lafdi K, Huie J (2007). Thermal conductivity of an aligned carbon nanotube array. Carbon, 45(13): 2608–2613.
112. Zhang K, Chai Y, Yuen MMF, Xiao DGW, Chan PCH (2008). Carbon nanotube thermal interface material for high-brightness light-emitting-diode cooling. Nanotechnology, 19(21): 215706.
113. Qiu L, Guo P, Kong Q, Tan CW, Liang K, Wei J, Tey JN, Feng YH, Zhang XX, Tay BK (2019). Coating-boosted interfacial thermal transport for carbon nanotube array nano-thermal interface materials. Carbon, 145: 725–733.
114. Kwak SY, Lew TTS, Sweeney CJ, Koman VB, Wong MH, Bohmert-Tatarev K, Snell KD, Seo JS, Chua NH, Strano MS (2019). Chloroplast-selective gene delivery and expression in planta using chitosan-complexed single-walled carbon nanotube carriers. Nat Nanotechnol, 14(5): 447–455.
115. Demirer G S, Zhang H, Goh NS, Pinals RL, Chang R, Landry MP (2020). Carbon nanocarriers deliver siRNA to intact plant cells for efficient gene knockdown. Sci Adv, 6 (26): eaaz0495.
116. Deshmukh MA, Jeon JY, Ha TJ (2020). Carbon nanotubes: An effective platform for biomedical electronics. Biosens Bioelectron, 150: 111919.
117. Zhu Z (2017). An overview of carbon nanotubes and graphene for biosensing applications. Nano-Micro Lett, 9(3): 1–24.
118. Dasari Shareena TP, McShan D, Dasmahapatra AK, Tchounwou PB (2018). A review on graphene-based nanomaterials in biomedical applications and risks in environment and health. Nano-Micro Lett, 10: 53.
119. Reina G, González-Domínguez JM, Criado A, Vázquez E, Bianco A, Prato M (2017). Promises, facts and challenges for graphene in biomedical applications. Chem Soc Rev, 46 (15): 4400–4416.
120. Cheng C, Li S, Thomas A, Kotov NA, Haag R (2017). Functional graphene nanomaterials based architectures: biointeractions, fabrications, and emerging biological applications. Chem Rev, 117(3): 1826–1914.
121. Fadeel B, Bussy C, Merino S, Vázquez E, Flahaut E, Mouchet F, Evariste L, Gauthier L, Koivisto AJ, Vogel U, Martín C, Delogu LG, Buerki-Thurnherr T, Wick P, Beloin-Saint-Pierre D, HischierR, Pelin M, Carniel FC, Tretiach M, Cesca F, Benfenati F, Scaini D, Ballerini L, Kostarelos K, Prato M, Bianco A (2018). Safety assessment of graphene-based materials: focus on human health and the environment. ACS Nano, 12(11): 10582–10620.

Chapter 2
Basic Physics of Carbon Nanotubes and Graphene

2.1 Basic Structure of Graphene and SWCNT

Structurally, carbon nanotubes (CNTs) are hollow cylinders formed by rolling up graphene sheets with different layers, thus graphene can be considered as a unit structure of CNTs. the basic structures of monolayer graphene sheets will be first introduced before understanding of CNT structures. Monolayer graphene is a single layer 2D honeycomb lattice plane, which is composed of a closely packed six-carbon rings with C–C bond length of 0.142 nm. The unit cell of graphene is characterized by the two vectors $\boldsymbol{a}_1$ and $\boldsymbol{a}_2$ with a length of 0.246 nm and an angle of 60°, where two carbon atoms are at the positions 1/3 ($\boldsymbol{a}_1$ + $\boldsymbol{a}_2$) and 2/3 ($\boldsymbol{a}_1$ + $\boldsymbol{a}_2$), respectively. In graphene, the planar sp^2-hybrid orbitals make carbon atoms bond with surrounding three carbon atoms through the σ-bonds, while each atom donates an unpaired electron to form π-orbital electrons. Therefore, the unique atomic structures enable graphene excellent physical and chemical properties.

As shown in Fig. 2.1, when a graphene sheet is rolled up along the chiral vector C = n$\boldsymbol{a}_1$ + m$\boldsymbol{a}_2$ (n and m are integers and $\boldsymbol{a}_1$ and $\boldsymbol{a}_2$ are lattice unit vectors) from the origin point (0, 0) to the point (n, m), forming the circumference of a SWCNT [1]. This circumferential vector $\boldsymbol{c}$ (or chiral vector) can be correctly defined and described by the chiral indices (n, m), and the diameters and chiral angles of all SWCNTs can also be uniquely identified. According to different chiral indices, SWCNTs can be divided into three categories: armchair, zigzag and chiral tubes. Among them, the chiral indices (n, n) and (n, 0) are called armchair and zigzag tubes, respectively.

For chiral tubes, the chiral angle θ is usually used to describe the direction of the circumferential vector, which is defined as the angle between $\boldsymbol{a}_1$ and $\boldsymbol{c}$ and can be calculated according the following formula:

$$\cos\theta = \frac{\boldsymbol{a}_1 * \boldsymbol{c}}{|\boldsymbol{a}_1| * |\boldsymbol{c}|} = \frac{2n+m}{2\sqrt{n^2+nm+m^2}} \tag{2.1}$$

Y. Su, *High-Performance Carbon-Based Optoelectronic Nanodevices*,
Springer Series in Materials Science 319,
https://doi.org/10.1007/978-981-16-5497-8_2

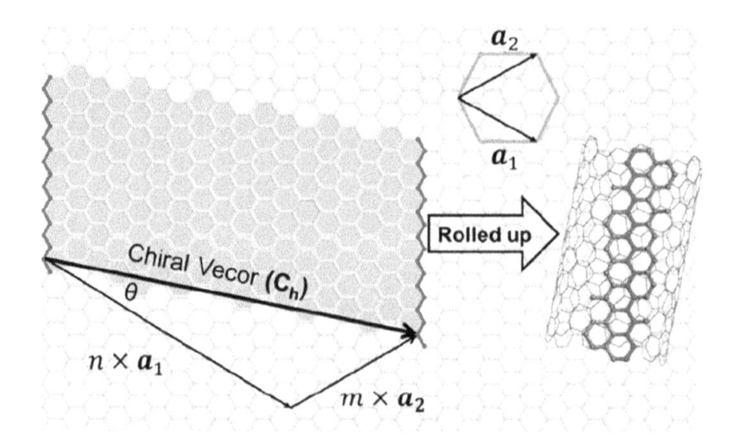

Fig. 2.1 Schematic diagram of graphene lattice and the corresponding structure of SWCNT with chiral indices (n, m). Reproduced with permission [1]. Copyright 2020, American Chemical Society

Owing to the six-fold rotational symmetry of graphene, the chiral angle θ in all SWCNTs is smaller than 60°. Meanwhile, each SWCNT with a chiral angle θ of 0°–30° has an equivalent tube with θ between 30°and 60°, except that the helix of graphene lattice around the tube is different. Therefore, the chiral angle θ is between 0° and 30° for all SWCNTs with n $\geq$ m $\geq$ 0, both armchair (θ = 30°) and zig-zag (θ = 0°) tubes are achiral tubes, and others are chiral tubes.

Except for chiral angle θ, the diameter, unit cell, atom number can be determined by the chiral indices (n, m), and the diameter of SWCNT can be calculated by the length of the chiral vector $\boldsymbol{c}$:

$$d = \frac{|\boldsymbol{c}|}{\pi} = \frac{a_0}{\pi}\sqrt{n^2 + nm + m^2} = 0.0783\sqrt{n^2 + nm + m^2}(nm) \tag{2.2}$$

where a_0 is the length of unit vector (0.246 nm). For the unit cell of tube, the translational period a along the axis of SWCNT is first defined by the smallest graphene lattice vector $\boldsymbol{a}$ perpendicular to $\boldsymbol{c}$, which can be expressed using the following formula [2].

$$\boldsymbol{a} = -\frac{2m+n}{NR}\boldsymbol{a}_1 + \frac{2m+n}{NR}\boldsymbol{a}_2 \tag{2.3}$$

$$a = |\boldsymbol{a}| = \frac{\sqrt{3(n^2 + nm + m^2)}}{NR}a_0 \tag{2.4}$$

where R = 3 if (n − m)/3 N is integer and R = 1 for otherwise. Thus, the unit cell of each SWCNT is formed by a cylindrical surface with circumference c and height a. Figure 2.2 shows the structures of (17,0), (10,10), and (12,8) tubes, it is obvious

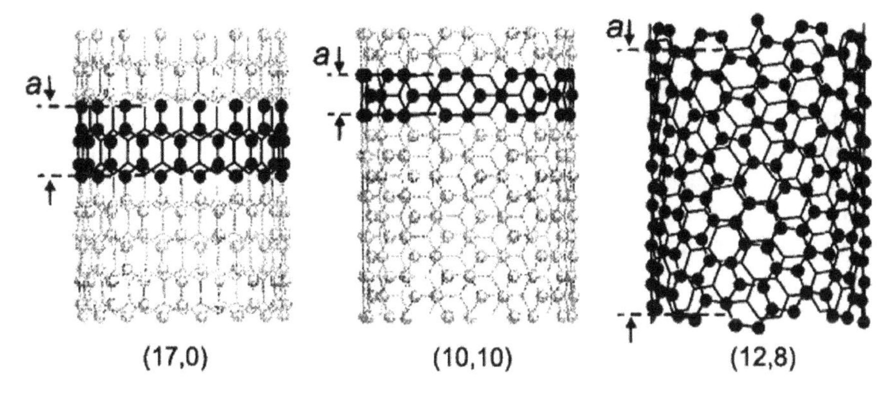

Fig. 2.2 Structure of the (17,0), the (10,0) and the (12,8) tube. The translational period a is indicated. Reproduced with permission [2]. Copyright 2007 John Wiley and Sons

that the translational period a is closely related the chirality of the tube, and chiral SWCNTs have longer unit cells than achiral tubes.

Since the unit cell of graphene contains two carbon atoms, the atom number (N_C) in the unit cell of SWCNT can be calculated from the ratio of the cylinder surface area (S_t) and the area S_g of the hexagonal graphene unit cell. The Nc number in the unit cell of SWCNT can be obtained as the following formula [2].

$$N_c = 2 \times \frac{S_t}{S_g} = \frac{4(n^2 + nm + m^2)}{N\mathcal{R}} \tag{2.5}$$

The structural parameters of armchair, zig-zag and chiral nanotubes are given summarized in Table 2.1.

2.2 Band Structure and Electronic Transport of Graphene

Before understanding the electronic band structure of SWCNTs, the band structure of graphene needs to be firstly introduced. As we know, the lattice structure of graphene consists of the hexagonal arrangement of carbon atoms, and three hybridized sp^2 orbitals in the same plane form the σ bonds between the adjacent

Table 2.1 Structural parameters of armchair, zig-zag and chiral nanotubes

SWCNT	M	θ	d	a	N_c
Armchair	$3n^2$	30°	$\sqrt{3}na_0/\pi$	a_0	4n
Zig-zag	n^2	0°	a_0/π	$\sqrt{3}a_0$	4n
Chiral	M	arcos $(n + m/2)/\sqrt{M}$	$\sqrt{M}a_0/\pi$	$\sqrt{3M}a_0/(NR)$	2M/(NR)

Note $M = n^2 + nm + m^2$

carbon atoms, while the vertical $2p_z$ orbital results in the π bonds out of the plane of graphene. The electronic band structure of graphene derived from π orbitals can be calculated by the tight-binding approximations. In a real space, a unit cell of graphene with unit vectors a_1 and a_2 is composed of two non-equivalent carbon atoms (A and B in Fig. 2.3a), while its reciprocal lattice with unit vectors b1 and b2 has three high-symmetry points (Γ, K and M in Fig. 2.3b). According to the tight-binding approximation, the energy dispersion of graphene in the reciprocal space can be obtained from the following formula: [3]

$$E = E_0 \mp \gamma_0 \sqrt{1 + 4\cos\left(\frac{\sqrt{3}k_x a}{2}\right)\cos\left(\frac{k_y a}{2}\right) + 4\cos^2\left(\frac{k_y a}{2}\right)} \tag{2.6}$$

where the negative and positive signs represent the valence bands of graphene formed by bonding π orbitals and the conduction bands formed by anti-bonding π* orbitals, respectively. The dispersion relation along h the high-symmetry Γ-K and Γ-M directions from the Eq. (2.6) when E_0 is zero is shown in Fig. 2.4a. The gap (∼11 eV) between the σ and σ* bands at the Γ point is so high that does not contribute to the common-used physical properties of graphene, it is usually neglected in the calculations. The π valence and π* conduction band cross at the K point of the Brillouin zone, and there are six such K points (Fig. 2.4b) at the corners of the 2D hexagonal Brillouin zone where the bandgap is zero [4]. The circular contour around each K point represents the conic shape of dispersion near the K point, and the slope of the conic shape near K points is determined by the Fermi velocity (8×10^5 m/s) of electrons in graphene. It is worth noting that only two K points in the reciprocal space are nonequivalent because they originate from two nonequivalent atoms (A and B) in the real space.

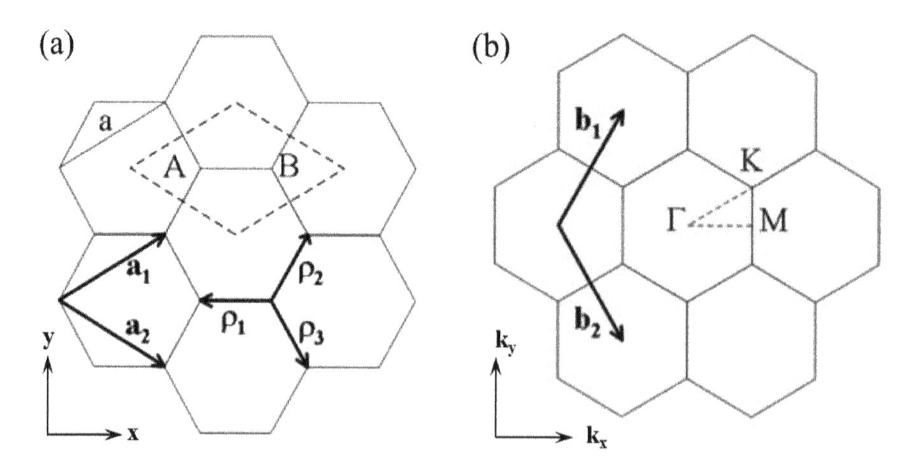

Fig. 2.3 **a** Real space and **b** reciprocal space schematic diagram of a graphene lattice

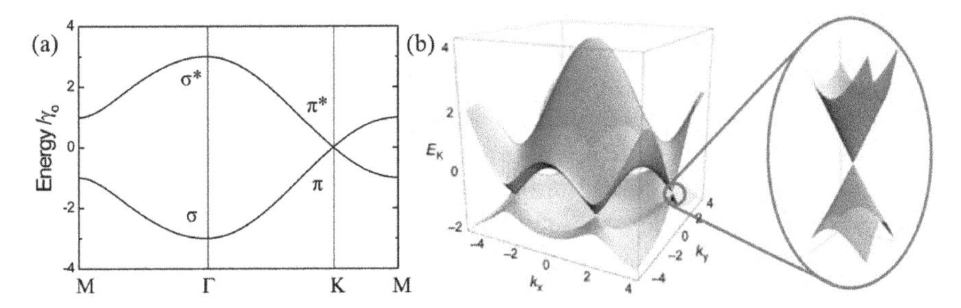

Fig. 2.4 **a** The energy dispersion of graphene along high-symmetry points with the Fermi level at zero. **b** The energy dispersion plotted as a function of wavevector. Reproduced with permission [4]. Copyright 2009 American Physical Society

From the energy dispersion relation of graphene with zero band gap, the density of states (DOS) is zero at the Fermi level, graphene is called a zero-gap semiconductor. The linear electronic dispersion near six K points makes the carriers in graphene available as massless Dirac fermions, and the six K points are referred to as the Dirac points where the nature of the carrier system can vary from electrons to holes in a single structure. These unique band structures are responsible for most notable electronic properties of graphene through the half-filled band that permits free-moving electrons. In addition, ideally 2D monolayer atomic structures enable graphene the substantial quantitative differences in the transport properties due to the carrier confinement effects. For example, the $2k_F$-backscattering in monolayer graphene is suppressed at low temperature.

2.3 Band Structure and Electronic Transport of SWCNTs

Since SWCNTs can be structurally considered to be a rolled-up sheet of graphene, its band structure can be obtained from that of graphene by introducing some periodic boundary conditions. Similar to graphene, the in-plane σ bonds are along the nanotube wall while the π bonds are perpendicular to the nanotube surface. Moreover, the out-plane π bonds also play a critical role in determining the electronic properties of SWCNTs. It is worth noting that the π orbitals are assumed to be still orthogonal to the σ orbitals when the band structure of SWCNTs is calculated only according to the π orbitals. This assumption may be no longer applicable for small SWCNTs due to high degree of hybridization between π and σ orbitals. On the other hand, high aspect ratio of SWCNTs makes its length regarded as infinite, the corresponding wave vector $k_{||}$ parallel to the nanotube axis is continuous, but the wave vector $k_{\perp}$ along the nanotube circumference must meet the periodic boundary condition, that is, the wave function should repeat itself after rotating 2π around the circumference, resulting in the quantized values of allowed $k_{\perp}$ for each SWCNT [3]. Therefore, the band structures of SWCNTs can be obtained by cross-sectional cutting the energy dispersion of graphene with the abovementioned allowed $k_\perp$ states, and several sub-bands can also be obtained by changing the cross-sectional cutting at different allowed $k_\perp$ states, as shown in Fig. 2.5a. As a result, the band structures of SWCNTs are closely related to the chiral indices of SWCNTs, which determine the spacing between discrete allowed $k_\perp$ states and the angles related to the surface Brillouin zone of graphene. For a sc-SWCNT, the allowed $k_\perp$ states do not pass through the K points as in Fig. 2.5b, the energy dispersion shows two parabolic bands with a chiral-indices related bandgap. Whereas they can pass directly through the K points (Fig. 2.5c) in a metallic SWCNT, resulting in two linear bands in the energy dispersion crossing at the Fermi level. Additionally, the band structures of SWCNTs can be modified by external factors, such atomic doping, molecule absorption, magnetic field, electric field, and mechanical deformation, and so on. These factors will influence the relative displacements between the allowed $k_\perp$ states and K points, resulting in a change of 1D band structure.

The 1D confinement in SWCNTs results in a quantization of the wave-vector perpendicular to the SWCNT axis. This leads to a characteristic distribution of DOS in k-space per energy level and the DOS is proportional to $1/E^{1/2}$ for each sub-band. The emerging sharp features in the distribution of DOS are so-called van Hove singularities, which represent the occupied (valence) and unoccupied (conduction) states. The DOS distribution of SWCNTs can be expressed as following [3, 5].

$$n(E) = \frac{2\sqrt{3}}{\pi^2}\frac{d}{\gamma_0 D}\sum_{m=-\infty}^{\infty} g(E, \varepsilon_m) \tag{2.7}$$

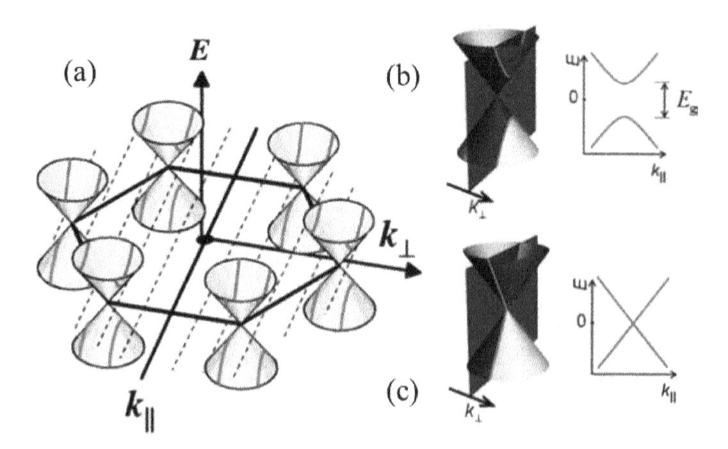

Fig. 2.5 **a** A first Brillouin zone of graphene with conic energy dispersions at six K points, and the allowed $k_\perp$ states in a SWCNT are presented by dashed lines. 1D energy dispersions for **b** a sc-SWCNT and **c** a semiconducting metallic SWCNT. Reproduced with permission [3]. Copyright 2008, Springer

where

$$g(E, \varepsilon_m) = \begin{cases} \frac{|E|}{\sqrt{E^2 - \varepsilon_m^2}}, & |E| > |\varepsilon_m| \\ 0, & |E| < |\varepsilon_m| \end{cases} \tag{2.8}$$

The van Hove singularity can be obtained when g(E, ε_m) becomes divergent for E = ε_m. The 1D DOS distribution of sc-SWCNTs is shown in Fig. 2.6, and the DOS of metallic SWCNT is found to be finite at the Fermi level due to its 1D characteristic. Along with zero bandgap, metallic SWCNTs as truly metallic, unlike graphene whose DOS is zero at K points. Moreover, the interband transition process (E_{ii}, i = 1, 2, 3, etc.) are closely related to the chiral indices (n, m) of SWCNTs. The relationship between abovementioned allowed interband transitions and the SWCNT diameter has been organized into the Kataura plot [6].

The 1D structure feature and small diameter of SWCNTs usually result in various quantum transport phenomena, and a finite number of sub-bands are demonstrated to participate in the electrical transport under a given bias voltage. The transport through these sub-bands can be well described by the Landauer formula, where one parabolic sub-band and a ballistic transport are considered in the 1D system. Under a given bias voltage V, the net current through the 1D system can expressed in the following formula: [3]

$$I = \Delta nev = \left(\frac{2eV}{hv}\right)ev = \frac{2e^2}{h}V \tag{2.9}$$

where Δn, e and v are the excess electron density, electron charge and charge velocity, respectively. Obviously, the net current is independent of carrier velocity and only depended on the voltage in an ideal 1D system. Consequently, the

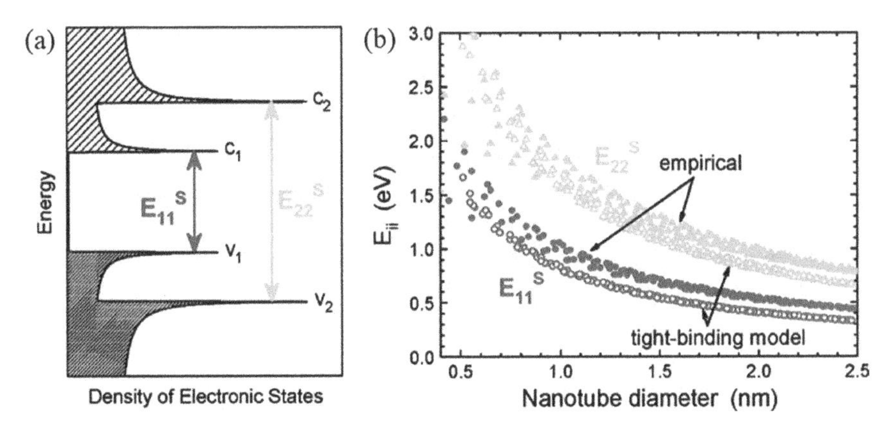

Fig. 2.6 **a** The electronic DOS distributions of sc-SWCNTs. **b** the Kataura plots via diameters. Reproduced with permission [6]. Copyright 2003, American Chemical Society

conductance quantum G_q and resistance quantum R_q in an ideal 1D system with one sub-band and no scattering can be obtained as follows:

$$G_q = \frac{I}{V} = \frac{2e^2}{h} \quad R_q = \frac{V}{I} = \frac{h}{2e^2} = 12.9\,k\Omega \tag{2.10}$$

For SWCNTs, two 1D sub-bands at the Fermi level participate in the electrical transport at low bias voltage, As a result, the resistance quantum R_Q is $\sim$6.5 kΩ [3].

The total current through multiple 1D channels is the net current per channel multiplied by the channel number *N*. If the electron scattering in each channel is taken into account, the total current I_T can be obtained after incorporating transmission probability (P_i) of electrons in each channel:

$$I_T = \frac{2e^2}{h} V \sum_i P_i \tag{2.11}$$

This is well known as the Landauer Formula, and $\sum P = N$ for ballistic transmission in an ideal 1D system.

Additionally, other quantum effects have been observed in the SWCNT system. For example, the Coulomb blockade effect has an important influence on electron transmission when a short SWCNT is coupled by tunneling barriers to two electrodes. If a long SWCNT has nearly ohmic contacts with the electrodes and single electron charging becomes negligible, it can be regarded as an ideal 1D quantum wire again. In this case, the interference of two propagating wave modes caused by the electron scattering at the SWCNT-metal interface plays a dominant role in the conductance of SWCNTs. As shown in Fig. 2.7, differential conductance maps as crisscross patterns reflects the interference of electron waves, also demonstrating the quantum mechanical wave nature of electrons and ballistic transport in SWCNTs [7, 8].

Although the ballistic transport properties through each 1D sub-band has been demonstrated in an ideal SWCNT system without any scattering, it is actually impossible to avoid the carrier scattering in the SWCNT sample, which directly influences the transport characteristics in SWCNT-based devices. In general, the carrier scattering can only happen at forward or backward direction in a SWCNT, which is quite different from those in 2D or 3D system. And the scattering in a SWCNT mainly originates from the static defects (structural defects and impurities) and phonons with time-depended potentials. The former can remarkably influences the carrier transport properties of sc-SWCNTs due to the sensitive change of conductance, but metallic SWCNTs have a long mean free path even at room temperature, this is because the backscattering in metallic SWCNTs is significantly suppressed [9, 10]. Additionally, the delocalized wave functions of SWCNTs near the Fermi level reduces the scattering effects, even in metallic SWCNTs containing defects [11]. Actually, it is very difficult to experimentally establish the correlations between the specific types of defects and the actual transport properties because the

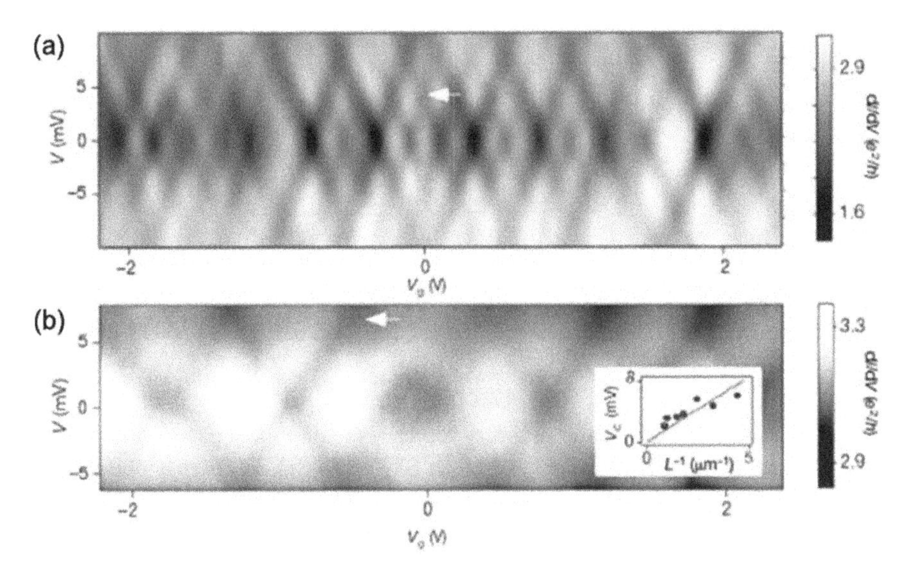

Fig. 2.7 The conductance maps in the quantum interference regime for a **a** 530 nm and **b** 220 nm SWCNT. Reprinted by permission [7]. Copyright 2001, Springer-Nature

number and types of structural defects cannot be precisely resolved for SWCNTs in the devices.

Since defect-induced scattering process is suppressed in metallic SWCNTs, the phonon scattering dominates, especially at high temperatures. This inelastic scattering process requires the momentum and energy conservations for the electron–phonon system, which allows only three possible electron–phonon backscattering processes in SWCNT to meet the requirement due to 1D structure characteristics and the symmetry, as shown in Fig. 2.8. The scattering process (Fig. 2.8a) induced by low-energy acoustic phonons involves a small momentum and energy changes. Therefore, this acoustic phonon scattering is the only available scattering process at room temperature and low electric fields due to the low energy of electrons, and the resistance is inversely proportional to the temperature [3]. For the optical and zone boundary phonon scattering in Fig. 2.8b, c, they require large energy changes and result in small and large momentum changes, respectively. Consequently, only the phonon emission induced by electrons happens for the two scattering process. However, high electric fields enable electrons enough energy to emit optical and zone boundary phonons, resulting in the backscattering of electrons [3]. For sc-SWCNTs, the main scattering mechanisms are also closely related the low- and high-energy phonons [12–14]. In a SWCNT FET, the scattering has been demonstrated to depend on the chirality and gate voltage due to the existence of additional sub-bands and gate voltage-controlled charge density in sc-SWCNTs. Meanwhile, another scattering from the metal contacts also needs to be considered [3, 14].

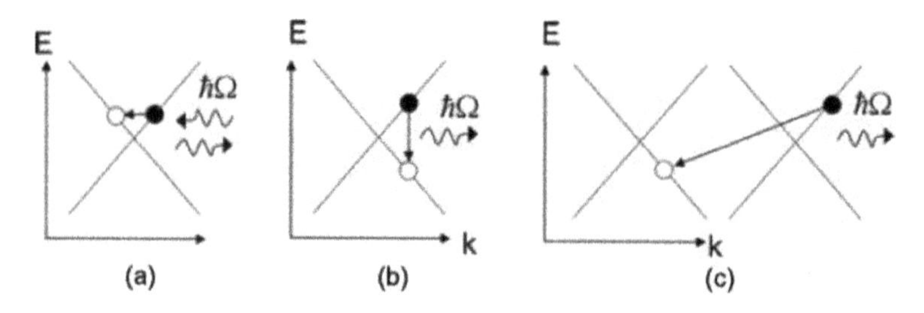

Fig. 2.8 Three allowed electron backscattering processes in a metallic SWCNT. **a** Acoustic phonons, **b** optical phonons, and **c** zone boundary phonons with energy ħΩ. Reprinted by permission [3]. Copyright 2008, Springer

For a sc-SWCNT, the carrier mobility is one of the important parameters that can characterize the transport properties of sc-SWCNTs. The carrier mobility in a channel determines the device performance of SWCNT FETs since it mainly reflects how fast the charge carriers respond to an external electric field. Typically, the carrier mobility of sc-SWCNTs refers to the effective and field-effect mobility because the Hall mobility is not suitable in the 1D SWCNT systems. The effective mobility (μ_{eff}) and field-effect mobility (μ_{FE}) are defined from the linear region of the transfer characteristics [15].

$$\mu_{\text{eff}} = \frac{LG}{C_g\left|V_g - V_t\right|}, \quad \mu_{\text{FE}} = \frac{L}{C_g}\left|\frac{\partial G}{\partial G_g}\right| \tag{2.12}$$

where L, G, C_g, V_g and V_t are the channel length, conductance, the gate capacitance per unit length, gate voltage, and the threshold voltage, respectively. From the transfer characteristic curves of MOSFET-like SWCNT-FETs, the experimentally reported mobility can reach highest value of $\sim 10^5$ cm^2/Vs [16]. Importantly, the μ_{FE} value of a sc-SWCNT FET can be estimated based on the band structure of sc-SWCNTs, and the G and μ_{FE} as a function of the gate voltage can be calculated from the characteristic curves of SWCNT FETs [15].

$$G = \frac{4l_0e^2\left(\Delta V_g/a\right)^2}{hL\left[1+\left(\Delta V_g/a\right)^2\right]}, \quad \mu_{\text{FE}} = \frac{e\tau_0\left(\Delta V_g/a\right)}{m^*\left[1+\left(\Delta V_g/a\right)^2\right]} \tag{2.13}$$

where $\Delta V_g = |V_g - V_t|$, $\alpha = 8e/3\pi dC_g$, $l_0 = v_0\tau_0$ is the mean free path related to the scattering time τ_0 at high energies. The G value increase in a gate voltage range determined by the parameter α, and then reaches saturation when the v_F approaches v_0 (8×10^5 m/s, the Fermi velocity in graphene). Similarly, the μ_{FE} value first increases linearly with ΔV_g and reaches a peak value of 0.32 $e\tau_0/m^*$, then decreases toward zero when the v_F approaches v_0 [15]. When the acoustic phonons dominates the scattering at small electric field and low temperature, the scattering time τ_0 is proportional to the SWCNT diameter and inversely proportional to the temperature. As a result, the peak mobility (μ_{peak}) and maximum conductance (G_{max}) can be expressed as following: [15]

$$G = \frac{4dv_0e^2}{\beta hLT}, \quad \mu_{peak} = \frac{0.48d^2v_0e}{\beta\hbar T} \tag{2.14}$$

where β is a proportionality coefficient, which can be extracted from either the maximum conductance or the peak mobility according to the experimental results. However, the effects of different interface contact and inhomogeneous in real FETs on the mobility need to be taken into account.

2.4 Optical Properties of SWCNTs

Under illuminations, the absorption coefficient (*a*) in SWCNTs can be directly given by the dielectric function with the following expression.

$$\alpha = \frac{2\kappa\omega}{c} = \frac{\varepsilon_2\omega}{nc} \tag{2.15}$$

where *c*, *K*, *n* and ε_2 are the light speed, extinction coefficient, refractive index and the imaginary part of the dielectric function, respectively. Interestingly, SWCNTs possess an exceptional nonlinear saturable absorption characteristic, that is, the optical absorption in SWCNTs does not saturate until the light intensity increases linearly to a given threshold value and then SWCNTs become transparent. Such saturable absorption feature makes SWCNTs act as a saturable absorber for passively mode-locked laser operation. In addition, SWCNTs have also been demonstrated to show exceptional high third-order optical nonlinearity, and the third order susceptibility $\chi^{(3)}$ is estimated theoretically to be in the order of $10^{-8}m^2/W$, which is considered to be responsible for many nonlinear processes, such as third harmonic generation, optical Kerr effect, phase conjugation, and so on.

The 1D band structures enable SWCNTs unique absorption and emission properties. For example, the photoexcited electron and hole pairs remain in an excitonic state with much larger binding energy due to the strong interaction with each other induced by the Coulomb interaction. Theses excitons sometimes dominate the light absorption and emission processes of SWCNTs based optoelectronic and photovoltaic devices. The exciton binding energy in SWCNTs is inversely proportional to the diameter and have a large binding energy in the order of 400 meV, it is usually difficult to become free carriers. The free electrons and holes can usually be produced in lower sub-bands (e.g. E_{11}) through two-step process as following. The electron-hole pairs are first excited to higher energy sub-bands (e.g. E_{22}), where the photoexcited carriers have shorter carrier lifetime ($\sim$130 fs) than those in lower sub-band ($\sim$1 ps), and then they relax very rapidly to the lower sub-band (e.g. E_{11}) to become free electrons and holes. Finally, these free electrons

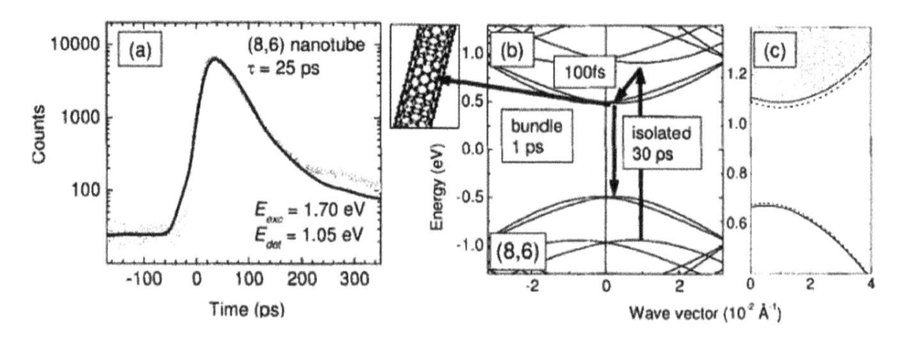

Fig. 2.9 Photoexcited carrier dynamics in SWCNTs. **a** Time-resolved photoluminescence measurement on an isolated (8, 6) nanotube. **b** Band structure of an (8, 6) nanotube and the carrier dynamics. Reproduced with permission [2]. Copyright 2007, John Wiley and Sons

and holes are extracted to other places or recombine again and return to the ground state with fluorescence. As shown in Fig. 2.9, for isolated SWCNTs, the electron and hole recombine across the first sub-band after 30 ps. much faster relaxation process happens in the SWCNT bundles in the order of picosecond time.

In order to investigate the luminescence properties of sc-SWCNTs, the isolated SWCNTs needs to be first obtained because the fluorescence quenching usually occurs in the SWCNT bundles due to non-radiative recombination [17]. In the photoluminescence (PL) measurement, high-energy photons corresponding to the second van Hove transition (E_{22}) are used to excite the electrons and hole to the second sub-band of sc-SWCNTs, and the recombination of photoexcited carriers emit strong fluorescence with low-energy photons corresponding to the first van Hove transition (E_{11}) after non-radiative relaxation. Moreover, the sharp absorption peaks in isolated SWCNTs can well match the photoluminescence emission peaks with no obvious shift, which is benefit for the precise assignment of chiral indices (n, m). In order to identify the chiral indices of the mixed SWCNTs, the PL excitation (PLE) spectrum mapping has been usually applied by plotting the PL intensity as a function of the energy of the exciting photons [18]. According to the energy relationship between the emitting photons and the excited photons, the roughly assignment can be achieved even if there are some apparent deviations between experimental results and those from theory calculation, which may be attributed to the curvature effect and the electron–hole interactions in the excited states.

2.5 Optical Properties of Graphene

Graphene also possesses exceptional optical properties due to the unique band structures, in which the interband optical transitions are allowed for all photon energies. Theoretically, the optical absorption of monolayer graphene is $\sim$2.3%

over the visible spectrum, [19] and the absorption coefficient of graphene is calculated to be as high as about 7×10^5 cm^{-1}, demonstrating the strong coupling effects between incident light and graphene. Moreover, the optical absorption is inversely proportional to the layer number of graphene. Importantly, the linear dispersion of the Dirac electrons in graphene indicates that an electron-hole pair is always formed simultaneously for any excitation. Ultrafast interband excitation will produce the non-equilibrium carriers in the valence and conduction band, and then these excited carriers decay very fast by different mechanisms. Generally, there are two type different relaxation processes with different timescales: The fast relaxation within 200–300 fs is usually associated with the carrier-carrier interactions (such as Auger and impact ionization) and phonon emission, another relaxation process takes place via optical phonon emission on a few picosecond timescale, which is usually attributed to the electron interband relaxation and cooling of hot phonons [20–22].

Another important optical properties of graphene is broadband saturable absorption and exceptional optical nonlinearity due to the Pauli blocking principle. Similar to SWCNTs, graphene is also a good saturable absorber material. Moreover, the linear dispersion of Dirac electrons makes the optical absorption independent of the wavelength, indicating that graphene as a saturable absorber can be used in the wider optical range extending from all telecommunications bandwidths to the mid- or far-infrared. Under illuminations with a low light intensity, the equilibrium between the formation and the recombination of photoexcited electron-hole pairs easily reach after photo-excitation due to the linear optical transition nature in graphene. However, the rapid increase in the concentration photoexcited electron–hole pairs results in a fact that the photoexcited carriers fill the valence bands when the incident light intensity is higher than a certain threshold, which blocks the further optical absorption and behave a transparent media to light. In addition, large third-order optical susceptibility is also found to be in mono- and multi-layer graphene [23, 24].

References

1. He M, Zhang S, Zhang J (2020). Horizontal single-walled carbon nanotube arrays: controlled synthesis, characterizations, and applications. Chem Rev, 120(22): 12592–12684.
2. Reich S, Thomsen C, Maultzsch J (2008). Carbon nanotubes: basic concepts and physical properties. John Wiley & Sons.
3. Kong J, Javey A (2009) Carbon Nanotube Electronics. New York, Springer-Verlag.
4. Neto AC, Guinea F, Peres NM, Novoselov KS, Geim AK (2009). The electronic properties of graphene. Rev Modern Phys, 81(1): 109.
5. Mintmire JW, White CT (1998). Universal density of states for carbon nanotubes. Phys Rev Lett, 81(12), 2506.
6. Weisman, RB, Bachilo, SM (2003). Dependence of optical transition energies on structure for single-walled carbon nanotubes in aqueous suspension: an empirical Kataura plot. Nano Lett, 3(9): 1235–1238.

7. Liang W, Bockrath M, Bozovic D, Hafner JH, Tinkham M, Park H (2001). Fabry-Perot interference in a nanotube electron waveguide. Nature, 411: 665–669.
8. Kong J, Yenilmez E, Tombler TW, Kim W, Dai H, Laughlin RB, Liu L, Jayanthi CS, Wu SY (2001). Quantum interference and ballistic transmission in nanotube electron waveguides. Phys Rev Lett, 87(10): 106801.
9. Ando T, Nakanishi T (1998). Impurity scattering in carbon nanotubes–absence of back scattering. Japan J Phys Soc, 67(5): 1704–1713.
10. White CT, Todorov TN (1998). Carbon nanotubes as long ballistic conductors. Nature, 393 (6682): 240–242.
11. Choi HJ, Ihm J, Louie SG, Cohen ML (2000). Defects, quasibound states, and quantum conductance in metallic carbon nanotubes. Phys Rev Lett, 84(13): 2917.
12. Pennington G, Goldsman N (2003). Semiclassical transport and phonon scattering of electrons in semiconducting carbon nanotubes. Phys Rev B, 68(4): 045426.
13. Perebeinos V, Tersoff J, Avouris P (2005). Electron-phonon interaction and transport in semiconducting carbon nano-tubes. Phys Rev Lett, 94(8): 086802.
14. Guo J, Lundstrom M (2005). Role of phonon scattering in carbon nanotube field-effect transistors. Appl Phys Lett, 86(19): 193103.
15. Zhou X, Park JY, Huang S, Liu J, McEuen PL (2005). Band structure, phonon scattering, and the performance limit of single-walled carbon nanotube transistors. Phys Rev Lett, 95(14): 146805.
16. Dürkop T, Getty SA, Cobas E, Fuhrer MS (2004). Extraordinary mobility in semiconducting carbon nanotubes. Nano Lett, 4(1): 35–39.
17. O'Connell MJ, Bachilo SM, Huffman CB, Moore VC, Strano MS, Haroz EH, Rialon KL, Boul PJ, Noon WH, Kittrell C, Ma JP, Hauge RH, Weisman RB, Smalley RE (2002). Band gap fluorescence from individual single-walled carbon nanotubes. Science, 297(5581): 593–596.
18. Bachilo SM, Strano MS, Kittrell C, Hauge RH, Smalley RE, Weisman RB (2002). Structure-assigned optical spectra of single-walled carbon nanotubes. Science, 298(5602): 2361–2366.
19. Nair RR, Blake P, Grigorenko AN, Novoselov KS, Booth TJ, Stauber T, Peres NMR, Geim AK (2008). Fine structure constant defines visual transparency of graphene. Science, 320(5881): 1308.
20. Kampfrath T, Perfetti L, Schapper F, Frischkorn C Wolf M (2005). Strongly coupled optical phonons in the ultrafast dynamics of the electronic energy and current relaxation in graphite. Phys. Rev. Lett., 95: 187s403.
21. Lazzeri M, Piscanec S, Mauri F, Ferrari AC, Robertson J (2005). Electronic transport and hot phonons in carbon nanotubes. Phys. Rev. Lett., 95: 236802
22. Bonaccorso F, Sun Z, Hasan TA, Ferrari AC (2010). Graphene photonics and optoelectronics. Nat Photonics, 4(9): 611.
23. Hendry E, Hale PJ, Moger J, Savchenko AK, Mikhailov SA (2010). Coherent nonlinear optical response of graphene. Phys Rev Lett, 105(9): 09740.
24. Bao Q, Loh KP (2012). Graphene photonics, plasmonics, and broadband optoelectronic devices. ACS Nano, 6(5): 3677–3694.

Chapter 3
Controlled Growths of Carbon Nanotubes and Graphene

3.1 Arc Discharge Synthesis of SWCNTs

As well known, CNTs were first observed from fullerene soot synthesized by direct current (DC) arc-discharge using transmission electron microscopy in 1991 [1]. Subsequently, SWCNTs also were successfully synthesized by this technique using magnetic transition metal (Fe, Co or Ni) as catalysts in 1993 [2, 3]. Since then, the arc discharge method has been one of the major methods for the preparation of high-quality SWCNTs. In the arc discharge process, a stable high-temperature plasma is produced in the gap between graphite cathode and fine graphite/catalyst anode in a low-pressure chamber by applying a certain current. Usually, the distance between two electrodes is 1–2 mm and remains stable during arc discharge process, and the chamber is filled by with a certain amount of inert gas. The extremely high temperature in the plasma region makes carbon and metal in the graphite/catalyst anode to be rapidly vaporized and the anode electrode is consumed constantly, which requires to continuously transferring the anode to keep a constant distance between two electrodes. The vaporized carbon and metal atoms are immediately carried away from the plasma zone by the insert gas and condensed into a liquid-phase solid solution. Subsequently, the carbon atoms in catalyst droplets precipitate and nucleate on the surface of catalysts at appropriate temperature and space domain, finally forming SWCNTs on the catalyst. Obviously, many parameters in the abovementioned process are very critical for the controlled synthesis of high-quality SWCNTs, Over the past three decades, the understanding and controlling the arc discharge process by these parameters, such as catalysts, buffer gas, current/voltage, electrode morphologies, carbon sources, and external fields has always attracted continuous attentions [4–8]. Firstly, metal catalyst is one of the most key parameters for the synthesis of SWCNTs, which is commonly expected to have low boiling temperature and high evaporation rate. In the first few years, different metal catalysts, including Ni, Fe, Co, Rh, and the related alloys, have been widely explored to synthesize the SWCNTs with different yields and

Y. Su, *High-Performance Carbon-Based Optoelectronic Nanodevices*,
Springer Series in Materials Science 319,
https://doi.org/10.1007/978-981-16-5497-8_3

diameters [3, 9, 10]. Among the metals, Fe and Ni have been demonstrated to be highly effective catalysts for the growth of high quality SWCNTs. Especially, the Fe and Ni/Y are most commonly used catalysts for the arc discharge synthesis of SWCNTs using S as a promoter. For example, Liu et al. [11] demonstrated a semi-continuous hydrogen arc discharge method using Fe/Co/Ni as catalysts and FeS as promoter, producing high-quality SWCNTs with a yield of ~ 2 g/h. Zhao et al. [12] synthesized high crystallinity SWCNTs with a purity higher than 70 at% by DC arc discharge with 1 at% Fe catalyst in H_2/Ar mixture gas, and the diameter of SWCNTs can be controlled by introducing W into Fe catalysts, which facilitates the synthesis of small-diameter SWCNTs [7]. For Ni-based catalysts, bi-or tri-metallic alloys have also been developed since first developed by Journet et al. [13], including Ni/Y, Ni/Mo, Ni/Co, Ni/Cr, and so on [13–16]. Among them, the Ni/Y catalysts have been widely used to commercially synthesize high-quality SWCNTs with the diameter of 1.2–1.7 nm under helium atmosphere. Meanwhile, the diameter and yield of SWCNTs can also be controlled by introducing Mo or S into Ni/Y catalysts [8, 17].

Since the most efficient catalysts to prepare SWCNTs is to combine Fe-based catalyst with H_2/Ar (or H_2) atmosphere or Ni-based catalysts with He atmosphere, the inset gas and pressure play an important role in the SWCNT growth. During the arc discharge process, the insert gas molecules can be ionized by high-temperature plasma, providing a steady flow of ions between cathode and anode. Meanwhile, the insert gas with a certain pressure facilitates the heat exchange and mass transfer in the chamber due to a suitable thermal conductivity. Shi et al. [18] demonstrated a strong effect of helium atmosphere on the SWCNT yield, and the yield increased with the increase of gas pressure. The optimum pressure of insert gas has been found to 300–700 Torr. However, high-yield synthesis of SWCNTs was possible even at low pressure of 100 Torr [19]. On the other hand, the SWCNT synthesis can be also controlled by varying the gas composition in the chamber. For example, Farhat et al. [20] demonstrated that the diameter of SWCNTs decreased with the increase of argon/helium ratio. However, the nucleation and growth of SWCNTs with small diameters are easily inhibited by introducing the reactive gas into the insert gas [6, 21]. Ando et al. [22] found that the yield of SWCNTs synthesized at the H_2/N_2 atmosphere was higher when compared to the H_2/Ar atmosphere, Similar H_2/X(X Ne, Ar, Kr, Xe) atmosphere have also been demonstrated to be suitable for mass-production of high-yield SWCNTs [23]. Remarkably, by replacing He with air, the SWCNTs with diameter of 1.29–1.62 nm has also been successfully synthesized at low-pressure air atmosphere with Ni/Y as catalysts, [21] where oxygen in air simultaneously etches carbon impurities and some SWCNTs and thereby the formed carbon oxides combining with N_2 act as are used as buffer gas.

Commonly, the anode diameter is smaller than that of cathode, which enables an effective vaporization of anode due to high current density and a nucleation of liquid carbon at low temperature surrounding the cathode region. Therefore, the relative size of two electrodes also has an impact on the nucleation and growth process of SWCNTs by changing the plasma temperature distribution [24]. Huang et al. [25] reported that the usage of a bowl-like cathode could significantly improve the SWCNT synthesis, obtaining high oxidation-resistant SWCNT products with lower level of structural defects. Su et al. [17] demonstrated that the length distribution and purity of SWCNTs could be effectively tuned by changing the diameter ratio of cathode/anode electrodes. The cathode with certain diameter facilitated the growth of high crystallinity SWCNTs with high purity and longer length. Except for cathode/anode electrodes, the temperature field surrounding the cathode region also influences the nucleation and growth process of SWCNs. Zhao et al. [26] demonstrated that the large-scale synthesis (>45 g/h) of SWCNTs by controlling the temperature inside the chamber during the arc discharge process. The atmosphere temperature not only strongly affected the yield of SWCNTs but also the purity and diameter distributions of SWCNT bundles. The optimum temperature was found to 600 °C, and higher temperature resulted in the decrease of yield.

More importantly, the arc plasma parameters (plasma density, electron temperature, etc.) play a critical role in the synthesis of SWCNTs, which directly affect the nucleation and growth processes of SWCNTs. It has been demonstrated that the plasma density and electron temperature can be significantly modified by changing the arc current and adding extra magnetic field in the arc plasma region [4, 5, 16, 27]. Especially, Magnetic field as an effective tool has been commonly applied to confine the discharge or tune the plasma density. Keidar et al. [4, 5] applied the axial magnetic field to confine the arc plasma during the arc discharge process. Experimental results suggested that the average length of SWCNTs increased by a factor of 2 using magnetic-field enhanced arc discharge compared with the conventional method, which is contributed to the increase of nanotube growth rate due to the increase of the plasma density after applying magnetic field. Meanwhile, Volotskova et al. [28] demonstrated that axial magnetic field could tailor the chirality distribution of SWCNTs by controlling the magnetic field strength, and found that the SWCNTs with small diameter (about 1 nm) significantly increased with the increase of magnetic field from 0.2 to 2 kg. Similarly, the diameter and deposition rate of SWCNTs can also be changed by introducing strong axial magnetic field [29].

Although axial magnetic field can be used to change the arc plasma parameters to some extent, it is difficult to significantly regulate the plasma parameters during the arc discharge process, Su et al. [8, 30] first developed an simple and efficient way to tailor the arc plasma parameters by applying a transverse magnetic field perpendicular to the electric field, which only a weak magnetic field (10–30 G) can effectively change the plasma parameters and spatial distribution. The magnetic field-induced diameter diameter-selective synthesis of SWCNTs has been demonstrated for bimetallic Fe/Mo and Ni/Y catalysts. Especially, when Fe/Mo was used as catalyst, most SWCNT web was preferentially deposited in the front region of oriented arc, whereas few products in the back region, as shown in Fig. 3.1a–b. Moreover, the diameter-selective efficiency was found to be closely related with the direct of magnetic field (Fig. 3.1c–f), In particular, the large-diameter SWCNTs can be effectively separated into the front region when a magnetic field with opposite direction was applied, while the SWCNTs with small diameter were enriched in the back region.

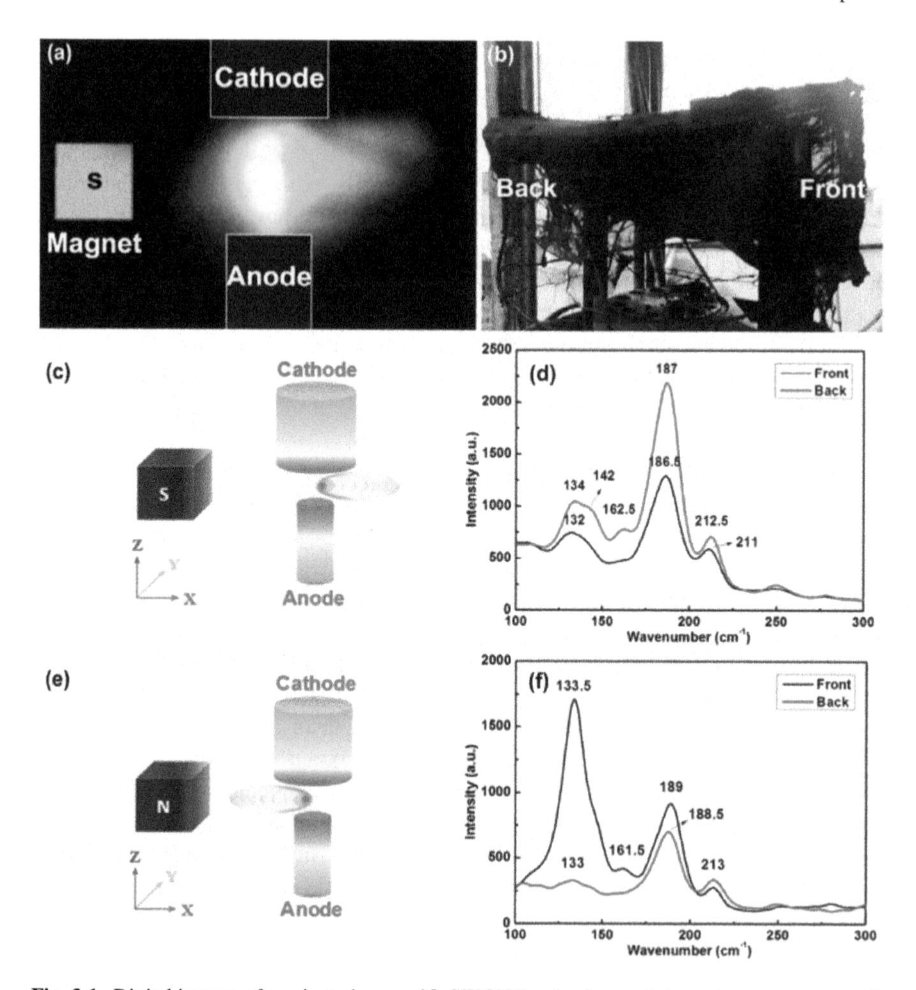

Fig. 3.1 Digital images of **a** oriented arc and **b** SWCNT web after applying a transverse magnetic field. **c** and **e** Schematic of the oriented arc under different magnetic field directions and corresponding Raman spectra (**d** and **f**) of SWCNTs collected in the front and back region of oriented arc. Reproduced with permission [30]. Copyright 2012, Elsevier

Over all, although enormous progress has been made in the large-scale and controlled synthesis of high quality SWCNTs by controlling the abovementioned factors, further experimental and theoretical investigations are still needed to establish a correlation between the SWCNT nucleation and the synthesis parameters, which is a key to better understand the growth mechanism SWCNTs. In the near future, it is expected to selectively synthesize the semiconducting and even specific-chiral SWCNTs by precisely tailoring the synthesis parameters.

3.2 Laser Ablation Synthesis of SWCNTs

Laser ablation method is also one of the high-temperature techniques extended from arc discharge method, which was first demonstrated by Guo et al. [31]. The basic principle can be briefly described as follow: a graphite containing metal catalyst is placed in a vacuum furnace with low-pressure inert gas at the high temperature condition, and high-energy pulsed or continuous laser beam irradiate the local region of the graphite, directly vaporizing the metal atoms and carbon atoms from the targeted graphite due to laser-induced ultrahigh temperature of 4000–6000 K. Then, the carrier gas bring the carbon–metal mixture in gas phase to the collector zone with appropriate temperature and space domain, the carbon atoms precipitate from the carbon–metal solid solution and nucleate on the surface of metal catalysts, the SWCNT growth happens on the collector. Compared with arc discharge method, laser ablation enables higher yield of SWCNTs with narrower diameter distribution and a relatively higher purity. The main reason is that the simultaneous evaporation of graphite and metal catalysts only happens in the local region where the laser is focused. Especially for pulsed laser, excess carbon species are rarely formed, thereby avoiding the formation of amorphous carbon or graphite particles, which is inevitable during the arc discharge process. However, this method requires high-energy density laser source and energy consumption, making it difficult to be commercially applied in the large-scale production of high-quality SWCNTs.

Similarly, the SWCNT synthesis by laser ablation technique is also influenced by many parameters such as the catalyst composition and contents, insert gas type and pressure, laser parameters, furnace temperature, and so on. Among them, metal catalysts play an critical role in the synthesis of SWCNTs, bimetallic Ni/Co or Ni/Y as highly efficient catalysts with a content of 1.2–9.0 at% have been commonly chosen during the laser ablation process [32–34]. Meanwhile, the insert gas pressure also has an impact on the local conditions near the evaporation zone, which plays an important role in the formation of SWCNTs. Munoz et al. [35] demonstrated that both argon and nitrogen atmospheres in the pressure range between 200 and 400 Torr facilitated high yield SWCNTs, whereas the formation of amorphous carbon was favored when the inert gas pressure was blow 200 Torr. Generally, the 200–500 Torr inert gas was considered to be a suitable pressure for the synthesis of SWCNTs [36, 37]. In addition, the formation process of SWCNTs is also influenced by the target and environment temperature, Lin et al. [34] reported the synthesis of SWCNTs via laser ablation using the targets containing acetates or nitrates in addition to Ni/Co catalysts. Experimental results confirmed that the mean diameter of SWCNTs showed a linear increase with temperature ranging from 1000 to 1200 °C. Moreover, the addition of nitrates improved the relative yield of SWCNTs over all the temperatures while the improved relative yield happened only at lower temperatures for acetates.

For laser ablation method, the laser parameters directly determine the conditions near the local temperature environment induced by the laser radiation, which is very critical for the nucleation and growth of SWCNTs [36, 37]. Dillon et al. [38]

demonstrated that the laser pulse energy could be used to tune the SWCNTs diameter from ~1.1 to 1.4 nm, and the diameters were shifted to smaller sizes for both room temperature and high temperature with increasing the pulse power. Zhang et al. [39] investigated the effect of laser power on the growth and diameter distribution of SWCNTs using continuous-wave CO_2 laser ablation without applying additional heat, and found that the average diameter of SWCNTs increased with increasing laser power, demonstrating the crucial role of laser power in the formation and diameter distribution of SWCNTs. Nowadays, different lasers or light sources have been developed to synthesize the SWCNTs, such as Nd:YAG lasers, UV lasers, continuous-wave CO_2 lasers, and solar energy [40]. For continuous-wave CO_2 laser, it does not require any external furnace in contrast to pulsed Nd:YAG laser system. However, taking advantage of the small heat affected zone of the pulse later region, the later can be used to precisely and selectively control the laser processes and synthesize SWCNTs at room temperature [41, 42]. Spellauge et al. [43] firstly achieved a pronounced debundling of the SWCNT network by operating the laser below the ablation threshold using femtosecond (470 fs) laser pulse ablation.

3.3 The CVD Synthesis of SWCNTs

The CVD method is the most studied and the main approach for the commercial production of SWCNTs due to suitable low temperature (i.e., 300–1200 °C), flexible parameter control, high production output efficiency and the cheaper cost. Dai et al. [44] first demonstrated the SWCNT synthesis through the disproportionation reaction of CO at 1200 °C using Mo particles as catalysts, and the synthesis temperature was further reduced to 1000 °C by choosing iron oxide as catalysts [45]. Nowadays, the transition metals have been used as effective catalysts to synthesize the SWCNTs with the best particle sizes of a few nanometers (typically 1–3 nm), and the optimized synthesis temperature is usually in the range of 700–900 °C. According to the supply type of catalysts, CVD can be simply divided into two kind of methods: floating-catalyst and catalyst-supported CVD. The floating-catalyst CVD is also called the aerosol synthesis or spray pyrolysis method, in which both catalysts and carbon source in gas phase are injected into a reactor at high temperature, thus the method is suitable for the low-cost synthesis of SWCNTs in large scale for practical application [46–49]. When the catalyst-supported CVD is utilized to synthesize SWCNTs, the catalyst precursors or solid catalyst nanoparticles need to be deposited on substrates first, and the catalyst sizes are usually controlled within 1–3 nm before SWCNT growth. Carbon source gases are subsequently introduced into the CVD chamber using Ar as carrier gas flow, and the subsequent decomposition, nucleation and growth processes are the same as those in the floating-catalyst CVD. In this section, we will introduce the abovementioned two type techniques in turn and their recent advances in the chirality-selective synthesis of SWCNTs.

3.3.1 The Floating-Catalyst CVD of SWCNT Powders

Among the common-used floating-catalyst CVD methods, high-pressure carbon monoxide (HiPco) gas-phase synthesis is the first commercial technique for mass production of SWCNTs. Nikolaev et al. [50] first reported high-yield catalytic synthesis of SWCNTs by thermally decomposing $Fe(CO)_5$ in a heated CO flow with the pressure of 1–10 atm at the temperatures of 800–1200 °C. The yield and diameter distribution of as-synthesized SWCNTs could be controlled by changing controlling the process parameters, resulting in the catalytic production of SWCNTs via the decomposition of $Fe(CO)_5$ in the presence of a continuous CO flow at high temperature and pressure. Subsequently, large-scale production (10 g/day) of high-purity SWCNTs had been demonstrated by further optimizing the process parameters including temperature, pressure, and catalyst concentration [51]. Nowadays, the HiPco gas-phase synthesis is still widely used to commercially synthesize the SWCNTs with small diameters.

Since then, different floating catalyst CVD methods have also been explored to synthesize SWCNTs using various carbon source and catalysts. For example, Cheng et al. [46] reported the floating-catalyst synthesis of SWCNTs using with ferrocene as catalyst precursor, in which benzene as the carbon source was introduced though H_2 flow as carrier gas. Nasibulin et al. [48] further developed the floating-catalyst CVD method using CO as the carbon source and demonstrated the important role of H_2 in the synthesis process. Marina et al. [49] reported the controlled synthesis of SWCNTs with a diameter distribution of 1.1–1.9 nm using CO as the carbon source and ferrocene as the catalyst. The increase of growth temperature could facilitate the formation of high quality and large-diameter SWCNTs, and the introduction of a small amount of CO_2 could enlarge the mean diameter due to the growth suppression of small-diameter nanotubes [52].

Except for CO as carbon sources, hydrocarbon gases have also been extensively utilized to synthesize SWCNTs based on floating-catalyst CVD method [53, 54]. For example, Hou et al. [55] demonstrated that small-diameter sc-SWCNTs and large-diameter metallic SWCNTs could be preferentially synthesized by tuning the species and amount of the carrier gas. Moreover, the introduction of H_2 served as an etching gas to remove the smaller-diameter sc-SWCNTs, resulting in the enrichment of large-diameter metallic SWCNTs. Hussain et al. [53] also synthesized long SWCNTs with controllable diameter-distribution using C_2H_4 gas as the only carbon source a carbon source and ferrocene as the catalyst precursor in an environmentally friendly and economical process. The mean diameter of the SWCNTs could be tuned in the range of 1.3–1.5 nm by controlling the ferrocene concentration, and its chirality distribution are between armchair and zigzag types. Meanwhile, the SWCNT synthesis can also be performed by injecting liquid carbon source containing catalyst precursor into a suitable reaction furnace [47, 56]. Ding et al. [56] reported high-quality SWCNT transparent conducting films by the floating-catalyst CVD method using ethanol as the carbon source and ferrocene as catalyst precursor. Experimental results illustrated that the mean diameter of SWCNTs increased with increasing the H_2 flow and most SWCNTs gathered near armchair nanotubes with 75–77% semiconducting types. Additionally, the effect of catalyst composition on the SWCNT growth was also investigated. The catalyst composition was found to efficiently tune the yield and characteristics of SWCNTs, but not dramatically change their chiralities [57]. The mean diameter of SWCNTs was mainly determined by the size distributions of catalyst particles, and the SWCNTs synthesized by Fe and Co catalysts had a wide chirality distribution from zig-zag to armchair edges, while the Co–Ni catalysts resulted in a narrower chirality distribution with 71% SWCNTs in the chiral angle of 15–30° [57].

Another effective floating-catalyst technique is the commercial-used fluidized-bed CVD, in which high surface/volume ratio of catalyst particles and high gas–solid mixing efficiency can be fully utilized as well as efficient heat transfer [58, 59]. Li et al. [60] successful synthesized SWCNTs using the fluidized-bed CVD using the nickel-nitrate coated silica-gel particles as catalysts and methane as carbon sources, in which the fluidization of a catalyst/support was involved by a hydrocarbon flow at high temperatures. The most critical part for the SWCNT synthesis using a fluidized bed is to design and synthesize the supported catalysts suitable for fluidization, various metal catalysts dispersed on the supported materials have been developed [59, 61]. Especially, Li et al. [62] recently developed high yield synthesis of submillimeter-long SWCNT arrays via fluidized-bed CVD using an Fe/AlO_x catalyst sputtered on ceramic beads, in which the moderately active C_2H_4 at 10–20 vol% and mildly oxidative CO_2 at 1 vol% was used to replace highly active C_2H_2 at low concentrations (0.3–1.1 vol%) and oxidative water vapor (ppm level). High-quality long SWCNTs with an average diameter of 2.9 nm and low catalyst impurity (0.1 mass%) have high specific surface area (1178 m^2/g) and high carbon yield of 28%. Such an effective technique provides a promising method for highly efficient mass production of long and pure SWCNTs.

It is worth noting that even through the abovementioned floating CVD techniques have been successful in the SWCNT synthesis, it is very challenging in synthesizing SWCNTs due to the existence of heterogeneous catalysts. Meanwhile, the aggregation of catalyst nanoparticles at high temperature usually results in the growth of large-diameter SWCNTs, which is unsuitable for controlling its chirality or diameter [63].

3.3.2 *The Catalyst-Supported CVD of SWCNTs*

As we know, the SWCNT powders synthesized by floating-catalyst CVD or high-temperature techniques need to be efficiently dispersed on the surface of supported substrates before the device fabrication. It not only introduces more uncontrollable factors, but also brings great challenges to fabrication of wafer-scale SWCNT-based devices. Therefore, the catalyst-supported CVD have been attracted an increasing attentions in the past two decades, in which the oxide buffer layers (such as SiO_2 and Al_2O_3) are usually utilized to prevent the reaction between the metal catalysts and substrate. Kong et al. [64] first demonstrated the synthesis of individual SWNTs on Si wafers patterned with micrometre-scale Fe catalyst using methane as carbon source. The as-synthesized SWCNTs have a diameter of 1–3 nm and a length of up to tens of micrometres, providing a promising approach to directly fabricate SWCNT-based devices on substrates. Since then, numerous efforts have been made to optimize the yield and diameter distributions of SWCNTs by changing catalyst types or sizes, carbon sources, process parameters, and other factors. The controlled synthesis of SWCNTs with narrow diameter and chirality distributions have been first achieved by tailoring the catalyst sizes and concentrations, buffer gas and other experimental parameters [65]. Then, the preferential growth of sc-SWCNTs or metallic SWCNTs were directly performed, significantly promoting the applications of SWCNTs in the novel nanoelectronics and photoelectric devices [66]. Li et al. [67] first developed the preferential synthesis sc-SWCNT networks at 600 °C using plasma enhanced CVD method, obtaining nearly 90% sc-SWCNTs randomly oriented on a substrate. However, the as-synthesized sc-SWCNTs still contain various chiralities because it is a considerable challenge to build the accurate correlation between nanotube chirality and catalysts. Therefore, the selective-growth of single-chirality SWCNTs with specific structure is always an ultimate goal for all synthesis techniques since 1993. Fortunately, Yang et al. [68] firstly demonstrated a direct the structure-specific growth of SWCNTs on SiO_2/Si substrates via ethanol CVD using W/Co bimetallic alloy nanocrystals as catalysts, selectively producing the (12, 6) SWCNTs with an abundance higher than 92%. The chirality-specific growth of (12, 6) SWCNTs was found to be attributed to good structural match between the atom arrangement around the circumference of (12, 6) SWCNTs and the arrangement of the catalytically active atoms in one of the planes of catalyst, as shown in Fig. 3.2. The research achievement provides a great possibility for direct growth of

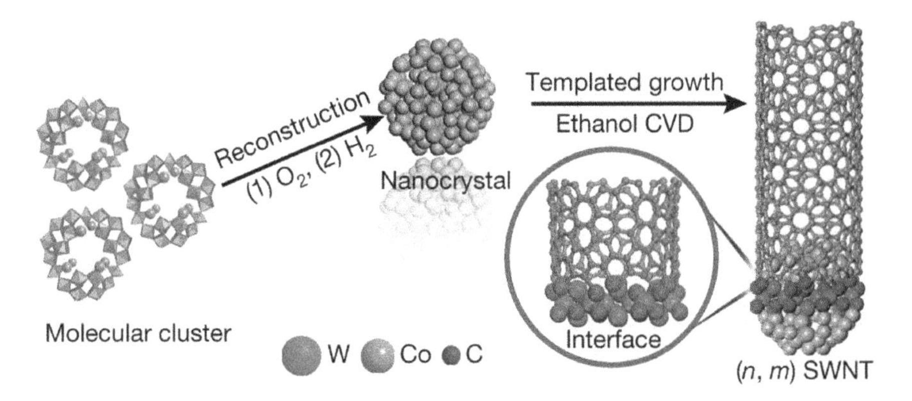

Fig. 3.2 Schematic diagram of the preparation of W/Co nanocrystal catalysts and corresponding specified (n, m) SWCNTs on the catalyst. Reproduced with permission [68]. Copyright 2014, Nature

single-chirality SWCNTs with chirality purity >99% by carefully controlling the composition, structure and size of the alloy catalysts and the process conditions.

As we know, the interactions between SWCNTs in random SWCNT networks will increase the electrical resistance to some extent, resulting in the limited performance of SWCNT-based nanoelectronics and photoelectric devices. Therefore, much more attentions have been attracted to the controlled synthesis of horizontally oriented SWCNTs on substrates, in which interconnection between two SWCNTs can be effectively avoided so that the electron transport in SWCNTs can be maximized between electrodes. Generally, in order to control the directions of SWCNT growth, different external forces have been applied during catalyst-supported CVD process, including gas flow, electric field, atomic step, crystal lattice, and so on. Meanwhile, similar to the floating-catalyst methods, the diameter-controlled synthesis of SWCNTs had been achieved at the initial stage by optimizing the catalysts, carbon sources and process parameters. The horizontally arranged sc-SWCNT array has been demonstrated to be directly synthesized by UV-assisted CVD, [69] producing over 95% sc-SWCNTs by destroying the metallic SWCNT cap in the beginning of growth. Ding et al. [70] demonstrated that metallic SWCNT could also be selectively etched by changing the growth atmospheres during the CVD process, producing over 95% sc-SWCNTs in the arrays. The preferential growth of sc-SWCNTs was mainly attributed to the introduction of methanol and ST-cut single crystal quartz as substrate. The OH radicals from methanol decomposition could selectively etch metallic SWCNTs, and the introduction of methanol also enhanced the SWCNT-substrate interaction. Similarly, the selective synthesis of sc-SWCNTs could also achieved by introducing water vapor into the isopropanol CVD system [71]. Moreover, the aligned SWCNT array successfully grown on quartz substrates up to 4 inches. The key factor in the process was the proper H_2O concentration and low carbon feed rate. To avoid the etching process during the selective synthesis of sc-SWCNTs, much more efforts have been made by designing the multi-period growth method or optimizing the catalysts, achieving highly efficient preferential growth of horizontally oriented sc-SWCNT arrays on substrates [72, 73].

With the understanding of the growth mechanism of sc-SWCNTs, numerous great advances have been achieved in the past few years. Especially, the solid-state catalysts exhibited a great promising potential for the chirality-selective growth of horizontally oriented sc-SWCNTs, since the specific interaction between nanotube and one of the planes of catalyst still remains fixed during the nucleate and growth process of SWCNTs. Zhang et al. [74] also demonstrated the chiral selective synthesis of horizontally oriented sc-SWCNTs using Mo_2C nanoparticles with uniform size (1.35 ± 0.2 nm) as solid state catalysts, which could selectively catalyze the scission of C–O bonds of ethanol molecules in the growth process. As a result, the metallic SWCNTs were preferentially etched by the absorbed oxygen (Oads), producing high-yield sc-SWCNTs with enriched (14, 4), (13, 6), and (10, 9) tubes. Meanwhile, the subnanometer-diameter SWCNT arrays with a narrower chiral distribution had also been synthesized using hydrogen-free CVD after optimizing the Mo_2C catalysts, resulting in the synthesis of SWCNT arrays with a density of $\sim$15 tubes/μm [75]. Importantly, higher chirality-selectivity in horizontally oriented SWCNT arrays had been successfully demonstrated on sapphire substrates by controlling the symmetries of the active catalyst surface and optimizing the growth parameters [76]. Metallic SWCNT arrays with an average density of >20 tubes/μm could be selectively synthesized using Mo_2C catalysts, in which the abundance of (12, 6) nanotubes was estimated to be greater than 90%. When WC catalysts were used, symmetry matching led to the growth of nanotubes with four-fold symmetry (Fig. 3.3a), such as (8, 4), (12, 6) and (16, 8) tubes. Especially, the (8, 4) tubes matched well with the (100) plane of WC catalyst in the symmetry, As a result, the abundance of (8, 4) nanotubes in SWCNT arrays with an average density of >10 tubes/μm was obtained. Similarly, horizontally oriented SWCNT arrays have also been developed using conventional catalysts in solid state by lowering the reaction temperature. For example, solid Co catalysts could be used to selectively synthesize horizontally oriented SWCNTs with different chirality at a temperature of 650 °C, Near-armchair (10, 9) tubes with $\sim$75% selectivity and the (12, 6) tubes with $\sim$82% have been successfully realized by choosing a small amount of ethanol and large amount of CO as carbon source, respectively [77]. Additionally, the (n, n − 1) family SWCNT arrays have also been selectively synthesized by carefully designing the solid catalyst and optimizing the growth conditions (Fig. 3.3b), [78] in which the near-equilibrium nucleation process of SWCNTs was explored by tuning the C/H ratios. Consequently, the abundances of 88% in (n, n − 1) SWNTs with large-diameter (>2 nm) and that >80% in single-chirality (10, 9) SWCNTs have been achieved, providing a new approach to control the structures of the SWCNT series.

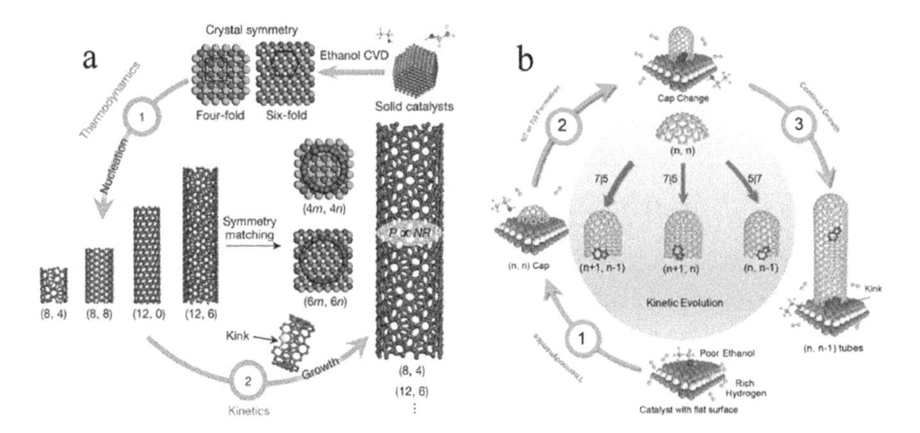

Fig. 3.3 **a** Schematic diagram of two-step chirality control of SWCNTs in ethanol CVD. Reproduced with permission [76]. Copyright 2017, Nature. **b** Schematic illustration of the mechanism of selective (n, n − 1) and (n, n − 2) SWCNT growth. Reproduced with permission [78]. Copyright 2019 Elsevier

Except for horizontally oriented SWCNT arrays, the vertically aligned SWCNTs forests have also attracted remarkable attentions due to its wide application prospect in various fields. Iijima et al. [79, 80] first demonstrated highly efficient synthesis of vertically aligned SWCNTs forests using water-assisted CVD method. The introduction of water vapor in the chamber was found to enhance significantly the activity and lifetime of the catalysts, resulting in superdense SWCNT forests with carbon purity of >99.98% and heights up to 2.5 mm (Fig. 3.4). More importantly, the approach provides a universal strategy to improve the mass production of SWCNTs synthesized by other synthesis methods. Since then, great achievements have been made by carefully designing the catalysts, changing carbon sources, and optimizing the growth conditions, thereby resulting in the improvement of synthesis efficiency and array density or the controlled synthesis of SWCNTs with certain diameter distributions [81–87]. However, the preferential growth of vertically aligned sc-SWCNTs is still a great challenge although the a high yield of up to 96% sc-SWCNTs have been successfully demonstrated by combining fast heating with plasma enhanced CVD [88].

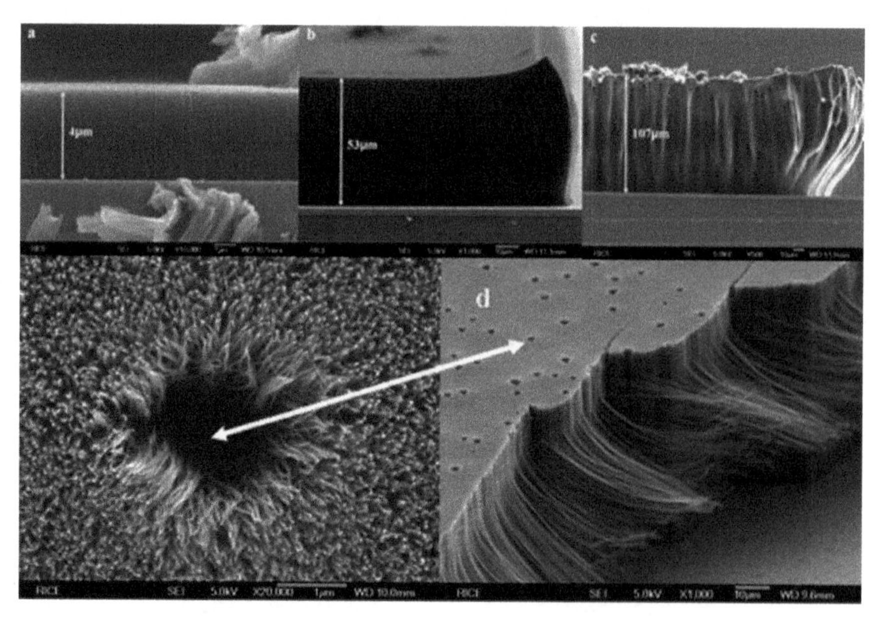

Fig. 3.4 SEM images of carpet SWCNTs grown with different time: **a** 1.3 min, **b** 20 min, **c** 40 min, and **d** inside view of the sample grown with 40 min. Reproduced with permission [82]. Copyright 2006, American Chemical Society

3.4 Catalytic CVD Growth of Graphene on Metal Substrates

Since the first report of graphene by mechanical peeling method in 2004, ddifferent synthesis techniques have been extensively investigated in the preparation of graphene based on physic or chemical mechanisms, such as mechanical or chemical exfoliation, epitaxial growth, catalyst-assisted or catalyst-free CVD growth, and so on. Although high-quality graphene by mechanical exfoliation meets the requirement of high performance graphene electronics and photoelectric devices, it is almost impossible to efficiently obtain large-scale graphene. Therefore, the CVD method is considered to be the most effective technique for the scalable preparation of high quality uniform graphene films due to its high controllability and scalability. Similar to the synthesis of SWCNTs, Ni and Co were chosen as catalyst to synthesize graphene at the early research works due to high carbon solubility and catalytic activity. After optimizing process parameters or growth conditions, high-quality graphene have been controllably synthesized. However, high carbon solubility in Ni and Co films make the precipitation rate and amount of dissolved C atoms difficult to be controlled, resulting in the non-uniform growth of graphene with inhomogeneous layer numbers. Therefore, other metal substrates with different carbon solubility and catalytic effect have also been used to synthesize graphene. Among them. Cu foils have been demonstrated the preferential growth of monolayer graphene due to suitable carbon solubility and the graphitization process only on the Cu surface, which was first demonstrated based on polycrystalline Cu foils as a substrate [89]. Since then, various Cu-based catalytic substrates have been widely utilized to synthesize large-area graphene films by controlling the nucleation and growth process on the substrates. Nowadays, large-area high-quality graphene films have been successfully prepared using sputtered Cu or Cu/Ni thin films, and polycrystalline Cu or Cu/Ni foils, single crystalline Cu or Cu/Ni foils [90]. However, the as-grown graphene films are polycrystalline consisting of small randomly rotated domains with a size of 500 nm to several μm, which significantly limits the physic performance of graphene-based electronic and photoelectric devices. Therefore, the large-scale CVD growth of single-crystalline graphene is highly required for high-performance electronics and other applications.

Importantly, the Cu(111) plane is found to be an ideal plane for the epitaxial growth of monolayer graphene due to the small lattice mismatch with graphene. Consequently, the large-size single crystal Cu(111) foils have been pre-prepared before graphene growth by the contact-free annealing of commercial Cu foils, and large-area monolayer graphene films have been commercially produced through a simple scale up of both single crystal Cu(111) foils and CVD furnace. Meanwhile, the Ni is usually introduced to modify the nucleation and growth process of graphene, enabling the fast growth of single crystal graphene films with monolayer or controlled layer number by changing the Ni content in Cu/Ni(111) alloy foils [91]. For example, Zhang et al. [92] developed a new approach to synthesize large-area single-crystal graphene using single crystal Cu/Ni(111) films/sapphire wafers

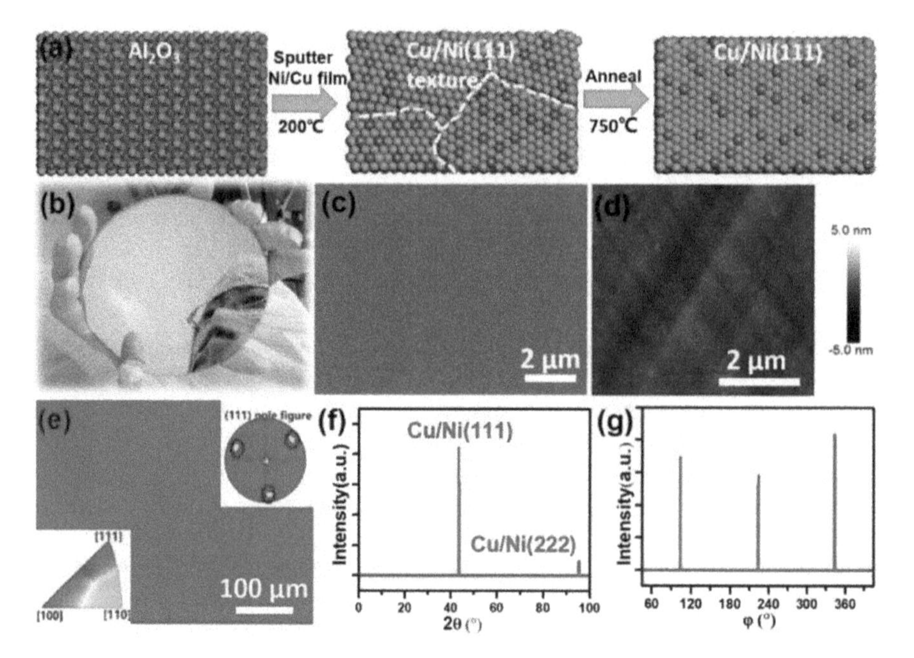

Fig. 3.5 Fabrication and characterizations of single-crystal Cu/Ni(111) film. Reproduced with permission [92]. Copyright 2019, Wiley

(c-plane, 6 in.) at lower temperature of 750 °C. As shown in Fig. 3.5, the Cu/Ni alloy film sputtered onto α-Al_2O_3(0001) wafer was converted into single-crystal Cu/Ni(111) film after the annealing treatment annealed at 750 °C in the flow of Ar/H_2. A wafer-scale (6 in.) and wrinkle-free single-layer graphene could be achieved after a 60 min growth at 750 °C. Similarly, single crystal $Cu_{90}Ni_{10}$(111) films on the surface of sapphire have also been obtained by combining two-step sputtering process and annealing process [93]. Moreover, the scalable growth of single crystal graphene films has been achieved in a vertical furnace with a large volume (Fig. 3.6).

Except for single crystal Cu/Ni alloy film, polycrystalline Cu/Ni foils have been successfully developed to synthesize single crystal graphene by carefully controlling the nucleation of the graphene. For example, Wu et al. [94] developed an efficient strategy for growing large-area single-crystalline graphene by locally feeding carbon precursor on an optimized polycrystalline $Cu_{85}Ni_{15}$ foils, achieving a single nucleus on the entire substrate and thus enabled the growth of $\sim$1.5-in. high-quality single-crystalline graphene in 2.5 h. Using similar approach, Vlassiouk et al. [95] also reported the growth of a $\sim$30 cm long single crystal monolayer graphene film on the polycrystalline Cu/Ni alloy substrate.

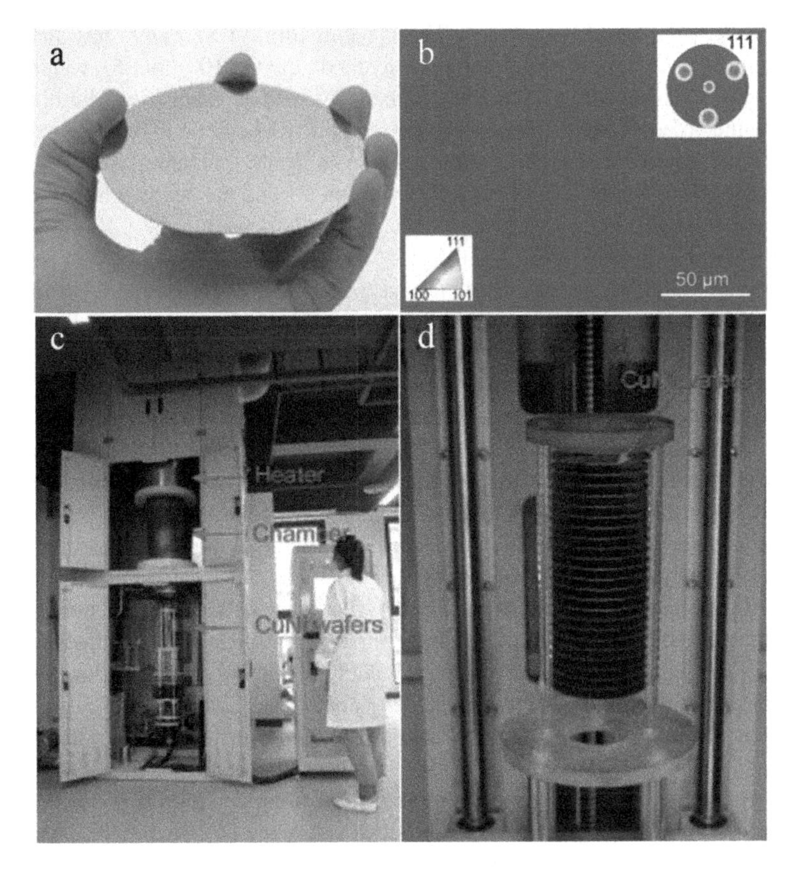

Fig. 3.6 **a** Photograph and **b** electron backscatter diffraction (EBSD) image of a 4 in. CuNi(111) thin film on sapphire. **c** A pilot-scale CVD furnace for scalable growth of single-crystal graphene film, and **d** 25 pieces of CuNi(111) wafer in a process cycle. Reproduced with permission [93]. Copyright 2019, Elsevier

3.5 Catalytic CVD Growth of Graphene on Nonmetal Substrates

Although the catalytic growth of graphene on the metal surfaces is the most effective approach to synthesize large-area, high-quality graphene films. the transfer process from metal substrates to the target substrates is unavoidable and the graphene quality is usually destroyed to some extent. Therefore, the direct growth of high quality graphene on the nonmetal substrates is highly desirable to avoid the costly, time-consuming and defect-inducing transfer process [96, 97]. Interestingly, semi-metal Ge has also been demonstrated the catalytic growth of graphene, which was firstly reported using ambient-pressure CVD by Di group [98]. Moreover, the crystal plane also plays a critical role in efficient growth of high quality graphene films, especially the (110) surface. The Ge thin film on Si wafers has also been successfully used to synthesize graphene layers on the Ge(100)/Si(100) wafers [99]. Moreover, it is concluded that the graphene growth is mainly self-limiting and surface-mediated due to the extremely low solubility of carbon in bulk Ge (<0.1%), thereby the growth temperature plays a key role in the graphene synthesis, which can determines the alignment of graphene fragments and the later agglomeration on Ge surfaces [100]. Recently, Li et al. [101] first demonstrate the CVD growth of single-crystal graphene film on the 15° miscut Ge(001) surface by perfectly aligning all the graphene islands, which has never been achieved on the normal Ge (001) surface. The suppression of graphene nucleation along the miscut direction of the vicinal surface is attributed to the unidirectional alignment of the graphene islands on the 15° miscut Ge(001) surface.

In addition, the catalytic-free growth of graphene on dielectric substrates has also attracted continuous interesting in the past decades. To assist the graphene growth on these substrates, external metal catalysts need to be introduced into the CVD growth system. For example, Ismach et al. [102] demonstrated the direct growth of graphene through surface catalytic decomposition of hydrocarbon precursors on thin Cu films pre-deposited on dielectric substrates. The catalytic Cu film was totally evaporated during or immediately after growth due to high temperature, resulting in direct deposition of graphene on different dielectric substrates without any post-etching process. In order to efficiently remove the metal residues, gas-phase metal promotors have been developed to catalyze the CVD growth of graphene. Kim et al. [103] demonstrated the efficient growth of high quality graphene by placing a suspended Cu foil on top of the insulator substrate without physical contact, in which a sufficient supply of homogenous Cu vapor could be provided during the graphene growth process, achieving high quality graphene that is comparable to that of graphene on the metal substrate. Remarkably, similar catalytic growth are also applicable to the growth of graphene on glass, achieving a large-area, uniform, and high-quality graphene glass with high transparency and conductivity (i.e., Rs = 370–510 $\Omega\ sq^{-1}$ at T = 82%) as well as uniform transmittance in the visible region [104]. And the flexible graphene glass has also be directly prepared using Cu foam-assisted plasma-enhanced CVD [105]. The photograph of the copper-foam-assisted grown flexible graphene glass revealed that both the transparency and conductivity of the macroscopic film.

3.6 Catalytic-Free CVD Growth of Graphene on Dielectric and Insulating Substrates

To completely avoid the effect of metal residues, the direct growth of graphene on insulating substrates is considered to be the most optimal approach by modifying the nucleation and growth process in the CVD process. Chen et al. [106] demonstrated the direct growth of graphene on the pre-treated SiO_2 surface using oxygen-aided CVD, in which the oxygen existing in the surface of SiO_2 could effectively promote the nucleation of graphene because oxygen facilitated the enhance adsorption of hydrocarbons on SiO_2. By replacing SiO_2 substrates by $SrTiO_3$ substrates with high catalytic ability, the direct growth of high quality graphene can also be realized at a relatively low temperature of 1050 °C [107]. Wei et al. [108] further reduced the growth temperature to 400 °C by introducing H_2 plasma into a graphene CVD growth system. It was found that the etching, critical edge growth and nucleation of graphene could be realized by changing the H_2 content and process parameters. By utilizing C_2H_4 as the carbon source, the micrometer-scale graphene crystals were grown directly on the SiO_2/Si substrates.

Additionally, since the nucleation and growth of graphene on the surface of melted Cu and Ga can be surprisingly better controlled than those on solid substrates, [109, 110] the liquid insulating substrates was expected to facilitate the uniform nucleation and speed up the growth of graphene during the direct CVD process [96]. Liu et al. [111, 112] first demonstrated the direct deposition of uniform, well-dispersed graphene disks and their continuous films on molten glasses, as shown in Fig. 3.7. Such directly grown graphene on low-cost glass substrates provide a novel approach to develop transparent conductive and extend the functions of typical glass production. Significantly, 12-in. uniform monolayer graphene film on molten glass has also directly growth on the soda-lime glass [113]. Nowadays, great achievements have been made and different growth techniques have also been well developed, such as direct thermal CVD growth, molten-bed CVD growth, metal-catalyst-assisted growth, and plasma-enhanced growth [96].

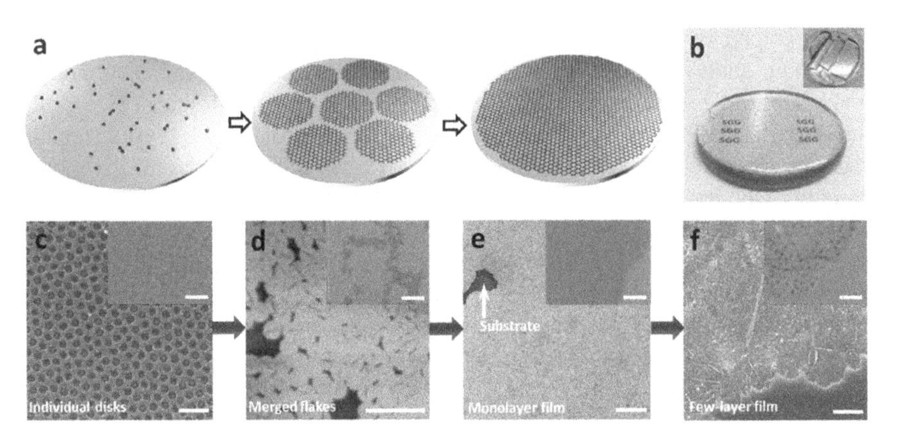

Fig. 3.7 **a** Schematic of direct graphene growth on molten glass. **b** Photograph of a graphene glass plate. **c–f** Morphology evolutions of as-grown samples experiencing different CVD synthetic conditions. Reproduced with permission [111]. Copyright 2015, John Wiley and Sons

3.7 Summary and Outlook

In summary, the common-used synthesis techniques and the state-of-the-art research progress of SWCNTs and graphene has been systemically summarized. Obviously, both the arc discharge and laser ablation techniques are still used to synthesize high-quality low-defect SWCNTs, but there is still a lack of breakthrough in the chirality-selective synthesis of SWCNTs. Although the intense evaporation of graphite and catalysts involved at high temperature is very difficult to be controlled efficiently, some parameters have been demonstrated to control the diameter distribution and yield of SWCNTs to some extent. It is worth noting that the phase changes and crystal orientation of supersaturated carbon-catalyst system have not received enough attention yet, because the nucleation and growth of SWCNTs are closely related to the catalysts under a certain temperature range. Therefore, it is necessary to pay more attention to the phase and crystal orientation of catalyst particles in the future. In contrast, the abovementioned factors have always been systemically investigated in the controlled growth of SWCNTs using CVD, and high purity sc-SWCNTs with single chirality have been enriched by designing the catalysts and growth parameters. Even though, some issues are still needed to be considered as following: (1) Understanding the growth kinetics and thermodynamics of SWCNTs with different chiralities remains limited using ex situ/in situ techniques, thereby the controlled nucleation and growth of SWCNTs with specific chirality is still a huge challenge. (2) The mass production of monochiral SWCNTs has not yet realized, which possesses identical electrical and optical properties for potential applications. (3) The density of horizontally aligned sc-SWCNTs with single chirality does not satisfy the demanded for SWCNT-based electronics (125 tubes μm^{-1}).

For graphene, various CVD techniques have been proved to synthesize effectively high-quality graphene films on different substrates, and large-area monolayer graphene films can be commercially provided based on Cu substrates. Especially, the achievements of wafer-scale large-area graphene on the single-crystal sapphire promotes its potential industrial applications. However, high-cost and the incompatibility with existing micro/naon process technology limits the fields of potential applications to some extent. Therefore, it is highly required to explore the controlled growth of wafer-scale single crystal graphene on the SiO_2/Si substrates, which is very critical for the development of graphene-based electronic and photoelectric devices for the potential commercial applications.

References

1. Iijima S (1991). Helical microtubules of graphitic carbon. Nature, 354(6348): 56–58.
2. Iijima S, Ichihashi T (1993). Single-shell carbon nanotubes of 1-nm diameter. Nature, 363 (6430): 603–605.

3. Bethune DS, Kiang CH, De Vries MS, Gorman G, Savoy R, Vazquez J, Beyers R (1993). Cobalt-catalysed growth of carbon nanotubes with single-atomic-layer walls. Nature, 363 (6430): 605–607.
4. Keidar M, Levchenko I, Arbel T (2018). Magnetic-field-enhanced synthesis of single-wall carbon nanotubes in arc discharge. J Appl Phys, 103(9): 094318.
5. Keidar M, Levchenko I, Arbel T, Alexander M, Waas AM, Ostrikov K (2008). Increasing the length of single-wall carbon nanotubes in a magnetically enhanced arc discharge. Appl Phys Lett, 92(4): 043129.
6. Su Y, Yang Z, Wei H, Kong Eirc, Zhang Y (2011). Synthesis of single-walled carbon nanotubes with selective diameter distributions using DC arc discharge under CO mixed atmosphere. Appl Surf Sci, 257(7): 3123–3127.
7. Fang L, Sheng L, An K, Yu L, Ren W, Ando Y, Zhao X (2013). Effect of adding W to Fe catalyst on the synthesis of SWCNTs by arc discharge. Physica E, 50: 116–121.
8. Su YJ, Zhang Y, Wei H, Zhang LL, Zhao J, Yang Z, Zhang YF (2012). Magnetic-field-induced diameter-selective synthesis of single-walled carbon nanotubes. Nanoscale, 4(5): 1717–1721.
9. Saito Y, Okuda M, Koyama T (1996). Carbon nanocapsules and single-wall nanotubes formed by arc evaporation. Surf Rev Lett, 3(1): 863–867.
10. Saito Y, Tani Y, Miyagawa N, Mitsushima K, Kasuya A, Nishina Y (1998). High yield of single wall carbon nanotubes by arc discharge using Rh-Pt mixed catalysts, Chem Phys Let, 294: 593–598.
11. Liu C, Cong H T, Li F, Tan P H, Cheng H M (1999). Semi-continuous synthesis of single-walled carbon nanotubes by a hydrogen arc discharge method. Carbon, 37(11): 1865–1868.
12. Zhao X, Inoue S, Jinno M, Suzuki T, Ando Y (2003). Macroscopic oriented web of single-wall carbon nanotubes. Chem Phys Lett, 373(3–4): 266–271.
13. Journet C, Maser W K, Bernier P, Loiseau A, DL Chapelle ML, Lee R, Fischer JE(1997). Large-scale production of single-walled carbon nanotubes by the electric-arc technique. Nature, 388(6644): 756–758.
14. Yudasaka M, Sensui N, Takizawa M, Bandow S, Ichihashi T, Iijima S (1999). Formation of single-wall carbon nanotubes catalyzed by Ni separating from Y in laser ablation or in arc discharge using a C target containing a NiY catalyst. Chem Phys Lett, 312(2–4): 155–160.
15. Shi Z, Lian Y, Zhou X, Gu Z, Zhang Y, Iijima S, Zhang S (1999). Mass production of single-wall carbon nanotubes by arc discharge method, Carbon, 37(9): 1449–1453.
16. He D, Zhao T, Liu Y, Zhu J, Yu G, Ge L (2007). The effect of electric current on the synthesis of single-walled carbon nanotubes by temperature controlled arc discharge. Diam. Relat Mater, 16: 1722–1726.
17. Su YJ, Zhang YZ, Wei H, Qian BJ, Yang Z, Zhang YF (2012). Length-controlled synthesis of single-walled carbon nanotubes by arc discharge with variable cathode diameters. Physica E, 44(7–8): 1548–1551.
18. Shi Z, Lian Y, Liao FH, Zhou X, Gu Z, Zhang Y, Zhang SL (2000). Large scale synthesis of single-wall carbon nanotubes by arc-discharge method. J Phys Chem Solid, 61(7): 1031–1036.
19. Park YS, Kim KS, Jeong HJ, Kim WS, Moon JM, An KH, Lee YH (2002). Low pressure synthesis of single-walled carbon nanotubes by arc discharge. Synthetic Metals, 126(2–3): 245–251.
20. Farhat S, Lamy de La Chapelle M, Loiseau A, Scott C D, Lefrant S, Journet C, Bernier, P (2001). Diameter control of single-walled carbon nanotubes using argon-helium mixture gases. J Chem Phys, 115(14): 6752–6759.
21. Su Y, Wei H, Li T, Geng HJ, Zhang YF (2014). Low-cost synthesis of single-walled carbon nanotubes by low-pressure air arc discharge. Mater Res Bull, 50: 23–25.
22. Ando Y, Zhao X, Inoue S, Suzuki T, Kadoya T (2005). Mass production of high-quality single-wall carbon nanotubes by H_2-N_2 arc discharge. Diamond Relat Mater, 14(3–7): 729–732.

23. Zhao X, Ohkohchi M, Inoue S, Suzuki T, Kadoya T, Ando Y (2006). Large-scale purification of single-wall carbon nanotubes prepared by electric arc discharge. Diamond Relat Mater, 15(4–8): 1098–1102.
24. Fetterman AJ, Raitses Y, Keidar M (2008). Enhanced ablation of small anodes in a carbon nanotube arc plasma. Carbon, 46(10): 1322–1326.
25. Huang H, Marie J, Kajiura H, Ata M (2002). Improved oxidation resistance of single-walled carbon nanotubes produced by arc discharge in a bowl-like cathode. Nano Lett, 2(10): 1117–1119.
26. Zhao T, Liu Y (2004). Large scale and high purity synthesis of single-walled carbon nanotubes by arc discharge at controlled temperature. Carbon, 42: 2765–2777.
27. Saha S, Page AJ (2016). The influence of magnetic moment on carbon nanotube nucleation. Carbon, 105: 136–143.
28. Volotskova O, Fagan JA, Huh JY, Phelan J, Frederick R, Keidar M (2010). Tailored distribution of single-wall carbon nanotubes from arc plasma synthesis using magnetic fields. ACS nano, 4(9): 5187–5192.
29. Yokomichi H, Ichihara M, Kishimoto N (2014). Magnetically induced changes in diameter and deposition rate of single-walled carbon nanotubes in arc discharge. Jpn J Appl Phys, 53 (2): 02BD05.
30. Su Y, Zhang YZ, Wei H, Yang Z, Kong ESW, Zhang YF (2012). Diameter-control of single-walled carbon nanotubes produced by magnetic field-assisted arc discharge. Carbon, 50(7): 2556–2562.
31. Guo T, Nikolaev P, Rinzler AG, Tomanek D, Colbert DT, Smalley RE (1995). Self-assembly of tubular fullerenes. J Phys Chem, 99(27): 10694–10697.
32. Yudasaka M, Komatsu T, Ichihashi T, Lijima S (1997). Single-wall carbon nanotube formation by laser ablation using double-targets of carbon and metal. Chem Phys Lett, 278 (1–3): 102–106.
33. Maser WK, Munoz E, Benito AM, De La Fuente GF, Maniette Y, Sauvajol JL (1998). Production of high-density single-walled nanotube material by a simple laser-ablation method. Chem Phys Lett, 292(4–6): 587–593.
34. Lin X, Rümmeli MH, Gemming T, Pichler T, Valentin D, Ruani G, Taliani C (2007). Single-wall carbon nanotubes prepared with different kinds of Ni-Co catalysts: Raman and optical spectrum analysis. Carbon, 45(1): 196–202.
35. Munoz E, Maser W K, Benito A M, De La F GF, Maniette Y, Righi A, Anglaret E Sauvajol JL (2000). Gas and pressure effects on the production of single-walled carbon nanotubes by laser ablation. Carbon, 38(10): 1445–1451.
36. Maser WK, Benito AM, Martınez MT (2002). Production of carbon nanotubes: the light approach. Carbon, 40(10): 1685–1695.
37. Munoz E, Maser WK, Benito AM, Martinez MT, De la Fuente GF, Righi A, Sauvajol JL, Anglaret E, Manieete Y (2000). Single-walled carbon nanotubes produced by cw CO_2-laser ablation: study of parameters important for their formation. Appl Phys A, 70(2): 145–151.
38. Dillon AC, Parilla PA, Alleman JL, Perkins JD, Heben MJ (2000). Controlling single-wall nanotube diameters with variation in laser pulse power. Chem Phys Lett, 316(1–2): 13–18.
39. Zhang H, Ding Y, Wu C, Chen YM, Zhu YJ, He YY, Zhong S (2003). The effect of laser power on the formation of carbon nanotubes prepared in CO_2 continuous wave laser ablation at room temperature. Physica B, 325: 224–229.
40. Chrzanowska J, Hoffman J, Małolepszy A, Mazurkiewica M, Kowalewski TA, Szymanski Z, Stobinski L (2015). Synthesis of carbon nanotubes by the laser ablation method: Effect of laser wavelength. Physica Status Solidi (b), 252(8): 1860–1867.
41. Dixit S, Singhal S, Vankar VD (2017). Size dependent Raman and absorption studies of single walled carbon nanotubes synthesized by pulse laser deposition at room temperature. Optical Mater, 72: 612–617.
42. Dixit S, Shukla AK (2018). Raman studies of single-walled carbon nanotubes synthesized by pulsed laser ablation at room temperature. Appl Phys A, 124(6): 1–6.

43. Spellauge M, Loghin FC, Sotrop J (2018). Ultra-short-pulse laser ablation and modification of fully sprayed single walled carbon nanotube networks. Carbon, 138: 234–242.
44. Dai H, Rinzler AG, Nikolaev P, Thess A, Colbert DT, Smalley RE (1996). Single-wall nanotubes produced by metal-catalyzed disproportionation of carbon monoxide. Chem Phys Lett, 260(3–4): 471–475.
45. Kong J, Cassell AM, Dai H (1998). Chemical vapor deposition of methane for single-walled carbon nanotubes. Chem Phys Lett, 292(4–6): 567–574.
46. Cheng HM, Li F, Su G, Pan HY, He LL, Sun X, Dresselhaus MS (1998). Large-scale and low-cost synthesis of single-walled carbon nanotubes by the catalytic pyrolysis of hydrocarbons. Appl Phys Lett, 72(25): 3282–3284.
47. Singh C, Shaffer MSP, Windle AH (2003). Production of controlled architectures of aligned carbon nanotubes by an injection chemical vapour deposition method. Carbon, 41(2): 359–368.
48. Nasibulin AG, Moisala A, Brown DP, Jiang H, Kauppinen EI (2005). A novel aerosol method for single walled carbon nanotube synthesis. Chem Phys Lett, 402(1–3): 227–232.
49. Tian Y, Timmermans MY, Partanen M, Nasibulin AG, Jiang H, Zhu Z, Kauppinen EI (2011). Growth of single-walled carbon nanotubes with controlled diameters and lengths by an aerosol method. Carbon, 49(14): 4636–4643.
50. Nikolaev P, Bronikowski MJ, Bradley RK, Rohmund F, Colbert DT, Smith KA, Smalley RE (1999). Gas-phase catalytic growth of single-walled carbon nanotubes from carbon monoxide. Chem Phys Lett, 313(1–2): 91–97.
51. Bronikowski MJ, Willis PA, Colbert DT, Smith KA, Smalley RE (2001). Gas-phase production of carbon single-walled nanotubes from carbon monoxide via the HiPco process: A parametric study. J Vacuum Sci Tech A, 19(4): 1800–1805.
52. Liao Y, Hussain A, Laiho P, Zhang Q, Tian Y, Wei N, Ding EX, Khan SA, Nguyen NN, Ahmad S, Kauppinen EI (2018). Tuning geometry of SWCNTs by CO_2 in floating catalyst CVD for high-performance transparent conductive films. Adv Mater Interfaces, 5(23): 1801209.
53. Hussain A, Liao Y, Zhang Q, Ding EX, Laiho P, Ahmad S, Wei N, Tian Y, Jiang H, Kauppinen EI (2018). Floating catalyst CVD synthesis of single walled carbon nanotubes from ethylene for high performance transparent electrodes. Nanoscale, 10(20): 9752–9759.
54. Yadav MD, Dasgupta K, Patwardhan AW, Kaushal A, Joshi JB (2019). Kinetic study of single-walled carbon nanotube synthesis by thermocatalytic decomposition of methane using floating catalyst chemical vapour deposition. Chem Eng Sci, 196: 91–103.
55. Hou PX, Li WS, Zhao SY, Li GX, Shi C, Liu C, Cheng HM (2014). Preparation of metallic single-wall carbon nanotubes by selective etching. ACS Nano, 8(7): 7156–7162.
56. Ding EX, Jiang H, Zhang Q, Tian Y, Laiho P, Hussain A, Liao Y, Wei N, Kauppinen EI (2017). Highly conductive and transparent single-walled carbon nanotube thin films from ethanol by floating catalyst chemical vapor deposition. Nanoscale, 9(44):17601–17609.
57. Ahmad S, Liao Y, Hussain A, Zhang Q, Ding EX, Jiang H, Kauppinen EI (2019). Systematic investigation of the catalyst composition effects on single-walled carbon nanotubes synthesis in floating-catalyst CVD. Carbon, 149: 318–327.
58. Wang Y, Wei F, Luo GH, Yu H, Gu GS (2002). The large-scale production of carbon nanotubes in a nano-agglomerate fluidized-bed reactor. Chem Phys Lett, 364 (5–6): 568–572.
59. Dasgupta, K, Joshi, JB, Banerjee S (2011). Fluidized bed synthesis of carbon nanotubes-A review. Chem Eng J, 171(3): 841–869.
60. Li Y. L, Kinloch IA, Shaffer MS, Geng J, Johnson B, Windle AH (2004). Synthesis of single-walled carbon nanotubes by a fluidized-bed method. Chem Phys Lett, 384(1–3): 98–102.
61. Zhao MQ, Zhang Q, Huang JQ, Nie JQ, Wei F (2010). Layered double hydroxides as catalysts for the efficient growth of high quality single-walled carbon nanotubes in a fluidized bed reactor. Carbon, 48(11): 3260–3270.

62. Li M, Hachiya S, Chen Z, Osawa T, Sugime H, Noda S (2021). Fluidized-bed production of 0.3 mm-long single-wall carbon nanotubes at 28% carbon yield with 0.1 mass% catalyst impurities using ethylene and carbon dioxide. Carbon, 182: 23–31.
63. Maruyama T (2018). Current status of single-walled carbon nanotube synthesis from metal catalysts by chemical vapor deposition. Mater Express, 8(1): 1–20.
64. Kong J, Soh HT, Cassell AM, Quate CF, Dai HJ (1998). Synthesis of individual single-walled carbon nanotubes on patterned silicon wafers. Nature, 395(6705): 878–881.
65. Min YS, Bae EJ, Oh BS, Kang D, Park W (2005). Low-temperature growth of single-walled carbon nanotubes by water plasma chemical vapor deposition. J Am Chem Soc, 127(36): 12498–12499.
66. Harutyunyan AR, Chen G, Paronyan TM, Pigos EM, Kuznetsov OA, Hewaparkrama K, Seung M, Zakharov D, Stach EA, Sumanasekera GU (2009). Preferential growth of single-walled carbon nanotubes with metallic conductivity. Science, 326(5949): 116–120.
67. Li Y, Mann, D, Rolandi M, Kim W, Ural A, Hung S, Javey A, Cao J, Wang D, Yenilmez E, Wang Q (2004). Preferential growth of semiconducting single-walled carbon nanotubes by a plasma enhanced CVD method. Nano Lett, 4(2): 317–321.
68. Yang F, Wang X, Zhang D, Yang J, Luo D, Xu ZW, Wei J, Wang JQ, Xu Z, Peng F, Li X, Li R, Li Y, Li M, Bai X, Ding F, Li Y (2014). Chirality-specific growth of single-walled carbon nanotubes on solid alloy catalysts. Nature, 510(7506): 522–524.
69. Hong G, Zhang B, Peng B, Zhang J, Choi WM, Choi J Y, Kim JM, Liu Z (2009). Direct growth of semiconducting single-walled carbon nanotube array. J Am Chem Soc, 131(41): 14642–14643.
70. Ding L, Tselev A, Wang J, Yuan DN, Chu HB, McNicholas TP, Li Y, Liu J (2009). Selective growth of well-aligned semiconducting single-walled carbon nanotubes. Nano Lett, 9(2): 800–805.
71. Che Y, Wang C, Liu J, Liu B, Lin X, Parker J, Beasley C, Wong H, Zhou C (2012). Selective synthesis and device applications of semiconducting single-walled carbon nanotubes using isopropyl alcohol as feedstock. ACS Nano, 6: 7454–7462.
72. Li J, Liu K, Liang S, Zhou W, Pierce M, Wang F, Peng L, Liu J (2014). Growth of high-density-aligned and semiconducting enriched single-walled carbon nanotubes: decoupling the conflict between density and selectivity. ACS Nano, 8: 554–562.
73. Zhang S, Hu Y, Wu J, Liu D, Kang L, Zhao Q, Zhang J (2015). Selective scission of C–O and C–C bonds in ethanol using bimetal catalysts for the preferential growth of semiconducting SWNT arrays. J Am Chem Soc, 137: 1012–1015.
74. Zhang S, Tong L, Hu Y, Kang L, Zhang J (2015). Diameter-specific growth of semiconducting SWNT arrays using uniform Mo_2C solid catalyst. J Am Chem Soc, 137: 8904–8907.
75. Kang L, Deng S, Zhang S, Li Q, Zhang J (2016). Selective growth of subnanometer diameter single-walled carbon nanotube arrays in hydrogen-free CVD. J Am Chem Soc, 138: 12723–12726.
76. Zhang S, Kang L, Wang X, Tong L, Yang L, Wang Z, Qi K, Deng S, Li Q, Bai X, Ding F (2017). Arrays of horizontal carbon nanotubes of controlled chirality grown using designed catalysts. Nature, 543(7644): 234–238.
77. Zhang S, Lin D, Liu W, Yu Y, Zhang J (2019). Growth of single-walled carbon nanotubes with different chirality on same solid cobalt catalysts at low temperature. Small, 15(46): 1903896.
78. Zhang S, Wang X, Yao F, He M, Lin D, Ma H, Sun Y, Zhao Q, Liu K, Ding F, Zhang J (2019). Controllable growth of (n, n-1) family of semiconducting carbon nanotubes. Chem, 5: 1182–1193.
79. Hata K, Futaba DN, Mizuno K, Namai T, Yumura M, Iijima S (2004). Water-assisted highly efficient synthesis of impurity-free single-walled carbon nanotubes. Science, 306(5700): 1362–1364.

80. Futaba DN, Hata K, Yamada T, Hiraoka T, Hayamizu Y, Kakudate Y, Tanaike O, Hatori H, Yumura M, Iijima S (2006). Shape-engineerable and highly densely packed single-walled carbon nanotubes and their application as super-capacitor electrodes. Nat Mater, 5(12): 987–994.
81. Zhong G, Iwasaki T, Robertson J, Kawarada H (2007). Growth kinetics of 0.5 cm vertically aligned single-walled carbon nanotubes. J Phys Chem B, 111(8): 1907–1910.
82. Xu YQ, Flor E, Kim MJ, Hamadani B, Schmidt H, Smalley RE, Hauge RH (2006). Vertical array growth of small diameter single-walled carbon nanotubes. J Am Chem Soc, 128(20), 6560–6561.
83. Zhong G, Warner JH, Fouquet M, Robertson AW, Chen B, Robertson J (2012). Growth of ultrahigh density single-walled carbon nanotube forests by improved catalyst design. ACS Nano, 6(4), 2893–2903.
84. Xiang R, Einarsson E, Murakami Y, Shiomi J, Chiashi S, Tang Z, Maruyama S (2012). Diameter modulation of vertically aligned single-walled carbon nanotubes. ACS Nano, 6(8): 7472–7479.
85. Chen G, Sakurai S, Yumura M, Hata K, Futaba, DN (2016). Highly pure, millimeter-tall, sub-2-nanometer diameter single-walled carbon nanotube forests. Carbon, 107: 433–439.
86. Han ZJ, Ostrikov K (2012). Uniform, dense arrays of vertically aligned, large-diameter single-walled carbon nanotubes. J Am Chem Soc, 134(13): 6018–6024.
87. Liu M, An H, Kumamoto A, Inoue T, Chiashi S, Xiang R, Maruyama S (2019). Efficient growth of vertically-aligned single-walled carbon nanotubes combining two unfavorable synthesis conditions. Carbon, 146: 413–419.
88. Qu L, Du F, Dai L (2008). Preferential syntheses of semiconducting vertically aligned single-walled carbon nanotubes for direct use in FETs. Nano Lett, 8(9): 2682–2687.
89. Li X, Cai W, An J, Kim SY, Nah J, Yang DX, Piner R, Velamakanni A, Jung I, Tutuc E (2009). Large-area synthesis of high-quality and uniform graphene films on copper foils. Science, 324(5932): 1312–1314.
90. Huang M, Ruoff RS (2020). Growth of single-layer and multilayer graphene on Cu/Ni alloy substrates. Acc Chem Res, 53(4): 800–811.
91. Huang M, Biswal M, Park HJ, Jin S, Qu D, Hong S, Zhu Z, Qiu L, Luo D, Liu X, Yang Z, Liu Z, Huang Y, Lim H, Yoo WJ, Ding F, Wang Y, Lee Z, Ruoff RS(2018). Highly oriented monolayer graphene grown on a Cu/Ni(111) alloy foil. ACS Nano 12: 6117– 6127.
92. Zhang X, Wu T, Jiang Q, Wang H, Zhu H, Chen Z, Jiang R, Niu T, Li Z, Zhang Y, Qiu Z, Yu G, Li A, Qiao S, Wang H, Yu Q, Xie X (2019). Epitaxial growth of 6 in. single-crystalline graphene on a Cu/Ni (111) film at 750 °C via chemical vapor deposition. Small, 15: 1805395.
93. Deng B., Xin Z, Xue R, Zhang S, Xu X, Gao J, Tang J, Qi Y, Wang Y, Zhao Y, Sun L, Wang H, Liu K, Rummeli MH, Weng LT, Luo Z, Tong L, Zhang, X, Xie C, Liu Z, Peng H (2019). Scalable and ultrafast epitaxial growth of single-crystal graphene wafers for electrically tunable liquid-crystal microlens arrays. Sci Bull, 64: 659–668.
94. Wu T, Zhang X, Yuan Q, Xue J, Lu G, Liu Z, Wang H, Wang H, Ding F, Yu Q, Xie X, Jiang M (2016). Fast growth of inch-sized single-crystalline graphene from a controlled single nucleus on Cu-Ni alloys. Nat Mater, 15: 43–47.
95. Vlassiouk IV, Stehle Y, Pudasaini PR, Unocic RR, Rack PD, Baddorf AP, Ivanov IN, Lavrik NV, List F, Gupta N, Bets KV, Yakobson BI, Smirnov SN (2018). Evolutionary selection growth of two-dimensional materials on polycrystalline substrates. Nat Mater, 17: 318–322.
96. Chen Z, Qi Y, Chen X, Zhang Y, Liu Z (2019). Direct CVD growth of graphene on traditional glass: methods and mechanisms. Adv Mater, 31(9): 1803639.
97. Liu B, Sun J, Liu Z (2020). Direct Growth of graphene over Insulators by Gaseous-Promotor-assisted CVD: Progress and Prospects. ChemNanoMat, 6(4): 483–492.
98. Wang G, Zhang M, Zhu Y, Ding G, Jiang D, Guo Q, Wang X (2013). Direct growth of graphene film on germanium substrate. Scientific Rep, 3(1): 1–6.

99. Pasternak I, Wesolowski M, Jozwik I, Lukosius M, Lupina G, Dabrowski P, Baranowski JM, Strupinski W (2016). Graphene growth on Ge (100)/Si (100) substrates by CVD method. Scientific Rep, 6(1): 1–7.
100. Dedkov Y, Voloshina E (2020). Epitaxial graphene/Ge interfaces: a minireview. Nanoscale, 12(21): 11416–11426.
101. Li P, Wei, W, Zhang M, Mei Y, Chu PK, Xie X, Yuan Q, Di Z (2020). Waer-scale growth of single-crystal graphene on vicinal Ge (001) substrate. Nano Today, 34: 100908.
102. Ismach A, Druzgalski C, Penwell S, Schwartzberg A, Zheng M, Javey A, Bokor J, Zhang Y (2010). Direct chemical vapor deposition of graphene on dielectric surfaces. Nano Lett, 10: 1542–1548.
103. Kim H, Song I, Park C, Son M, Hong M, Kim Y, Choi HC (2013). Copper-vapor-assisted chemical vapor deposition for high-quality and metal-free single-layer graphene on amorphous SiO_2 substrate. ACS Nano, 7(8): 6575–6582.
104. Sun J, Chen Z, Yuan L, Chen Y, Ning J, Liu S, Ma D, Song X, Priydarshi M, Bachmatiuk A, Rümmeli M, Ma T, Zhi L, Huang L, Zhang Y, Liu Z (2016). Direct chemical-vapor-deposition-fabricated, large-scale graphene glass with high carrier mobility and uniformity for touch panel applications. ACS Nano, 10(12): 11136–11144.
105. Wei N, Li Q, Cong S, Ci H, Song Y, Yang Q, Lu C, Li C, Zou G, Sun J, Zhang Y, Liu Z (2019). Direct synthesis of flexible graphene glass with macroscopic uniformity enabled by copper-foam-assisted PECVD. J Mater Chem A, 7(9): 4813–4822.
106. Chen J, Wen Y, Guo Y, Wu B, Huang L, Xue Y, Geng D, Wang D, Yu G, Liu Y (2011). Oxygen-aided synthesis of polycrystalline graphene on silicon dioxide substrates. J Am Chem Soc, 133: 17548–17551.
107. Sun J, Gao T, Song X, Zhao Y, Lin Y, Wang H, Ma D, Chen Y, Xiang W, Wang J, Zhang Y, Liu Z (2014). Critical crystal growth of graphene on dielectric substrates at low temperature for electronic devices. J Am Chem Soc, 136: 6574–6577.
108. Wei D, Lu Y, Han C, Niu T, Chen W, Wee ATS (2013). Critical crystal growth of graphene on dielectric substrates at low temperature for electronic devices. Angew Chem Int Ed, 52: 14121–14126.
109. Geng D, Wu B, Guo Y, Huang L, Xue Y, Chen J, Yu G, Jiang L, Hu W, Liu Y (2012). Uniform hexagonal graphene flakes and films grown on liquid copper surface. Proc Natl Acad Sci, 109(21): 7992–7996.
110. Wu YA, Fan Y, Speller S, Creeth GL, Sadowski JT, He K, Robertson AW, Allen CS, Warner JH (2012). Large single crystals of graphene on melted copper using chemical vapor deposition. ACS Nano, 6(6): 5010–5017.
111. Chen Y, Sun J, Gao J, Du F, Han Q, Nie Y, Chen Z, Bachmatiuk A, Priydarshi MK, Ma D, Song X, Wu X, Xiong C, Rümmeli MH, Ding F, Zhang Y, Liu Z (2015). Growing uniform graphene disks and films on molten glass for heating devices and cell culture. Adv Mater, 27: 7839–7846.
112. Sun J, Chen Y, Priydarshi MK, Chen Z, Bachmatiuk A, Zou Z, Chen Z, Song X, Gao Y, Rümmeli MH, Zhang Y, Liu Z (2015). Direct chemical vapor deposition-derived graphene glasses targeting wide ranged applications. Nano Lett, 15: 5846–5854.
113. Chen Z, Ci H, Tan Z, Dou Z, Chen X, Li B, Liu R, Lin L, Cui L, Gao P, Peng H, Zhang Y, Liu Z (2019). Growth of 12-inch uniform monolayer graphene film on molten glass and its application in PbI2-based photodetector. Nano Res, 12: 1888–1893.

Chapter 4
Characterizations of Carbon Nanotubes and Graphene

4.1 Introduction

The excellent physical and chemical properties of carbon nanotubes (CNTs) and graphene are determined by their unique atomic structures. The comprehensive characterizations are needed to establish the relationships between the properties and unique structures. As we know, a series of structural and electronic parameters can be determined from the chiral indices of single-walled CNTs (SWCNTs): such as the diameter, chiral angle, conductive property, bandgap, interband optical transitions, and so on. Therefore, the indentification of chiral indices has been attracted much attention since the discovery of CNTs [1, 2]. Especially, the accurate structural characterizations are required during the selective growth and chirality separation of SWCNTs, [3–5] thereby obtaining detailed diameter and chirality distributions in the sample. For SWCNT-based electronic and optoelectronic devices, the band gap and purity of semiconducting SWCNTs need to be statistically indentified using optical spectroscopy techniques [6, 7]. For graphene, monolayer graphene is considered to be a zero-bandgap semiconductor, and rhombohedral-stacked trilayer graphene has been demonstrated to be semiconducting with a tunable band gap [8, 9]. While the twisted bilayer graphene becomes a superconductorat the magic angle of exactly 1.1° [10, 11], which is capable of carrying the electric current with no resistance. On the other hand, graphene nanoribbons exhibit a semiconducting property after cutting graphene into one-dimensional sturctures, in which zigzag and armchair edges influence their physical properties. Furthermore, the characterizations of pentagon-heptagon defects and their density in graphene are also very important, which provide an atomic-level strucutural information at the grain boundary and in the interior of polycrystalline grapheme [12, 13].

In this chapter, several typical structural characterization techniques have been systemically introduced, including Raman spectroscopy, electron microscopy, absorption spectroscopy and luminescence spectroscopy, and so on. Taking into

Y. Su, *High-Performance Carbon-Based Optoelectronic Nanodevices*,
Springer Series in Materials Science 319,
https://doi.org/10.1007/978-981-16-5497-8_4

consideration the characterization accuracy and efficiency, the collaborative utilization of different characterization technologies is recommended.

4.2 Raman Spectroscopy

4.2.1 SWCNTs

Raman spectroscopy has been commonly considered as the most convenient and powerful technique for characterizing individual or bulk SWCNTs without contact or destruction, [14] because the interactions between SWCNTs and incident light result in the frequency (or wavenumber) change of the scattered light due to the vibration and rotation of C–C bond. As we know, the interband transitions (E_{ii}, i = 1, 2, 3, 4, etc.) between van Hove singularities in density of electronic states is strongly depended on the diameter of SWCNTs due to its unique 1D structural feature, which determines the optical absorption process in SWCNTs. When the E_{ii} energy is close to the energy (E_{laser}) of incident or scattered light, the Raman scattering efficiency can be increased by a factor of approximately 10^3, that is, a resonance Raman scattering happens. Moreover, the smaller the difference between E_{ii} and E_{laser}, the higher the intensity of Raman scattering. At this time, even the Raman singal from individual SWCNT is still enough to be detected. Threfore, the resonance Raman spectroscopy can be used to investigate the electronic and phonon structures of individual SWCNTs [15–17].

Generally, four characteristic peaks can be observed from a typical Raman spectrum of SWCNTs: (a) **Radial breathing mode** (RBM) located at low-frequency region (100–300 cm^{-1}): originating from the coherent vibrations of the carbon atom in the radial direction (Fig. 4.1). The RBM peaks can be observed from Raman spectra of SWCNTs, double- and triple-walled CNTs, providing the information about diameter, conductive type and chirality of SWCNTs. (b) **D-band** with a wavenumber around 1350 cm^{-1}; arising from the structural defects and tube curvature. The peak position of D-band is also related to the diameter or chirality of SWCNTs to some extent. (c) **G-band** located at high-frequency region (1500–1600 cm^{-1}): originating from longitudinal optic (L_O) and in-plane tangential optic (i_{TO}) modes of phonons. G-band consists of radial-vibration related G^- band and axial-vibration related G^+ band ($\sim$1591 cm^{-1}): thereby, G^+ band is sensitive to the charge transfer between SWCNTs and others while the G^- band is closely depended on the conductive-type [18]. (d) **G′ (2D) band** with a wavenumber around 2700 cm^{-1}, resulting from the second order resonance Raman process. Similar to the D-band, G′-band is also sensitive to the diameter or conductive type of SWCNTs. In addition, for a metallic SWCNT (m-SWCNT): a new electronic Raman scattering (ERS) peak can be found except for the abovementioned characteristic peaks [19].

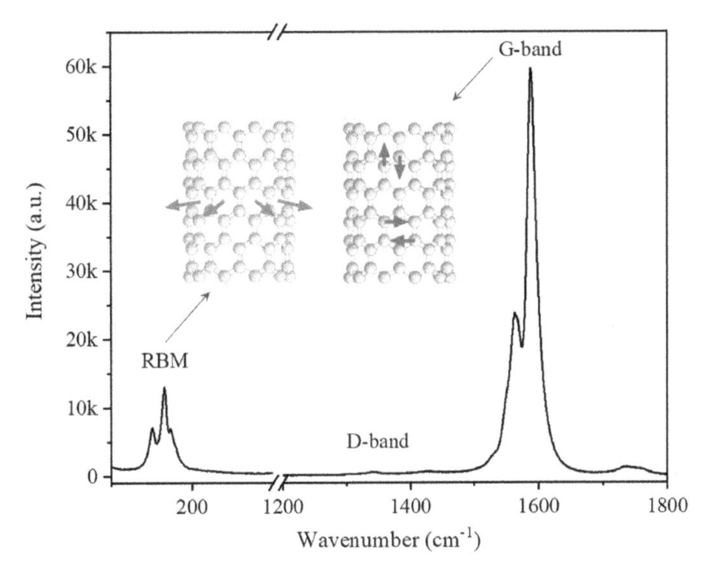

Fig. 4.1 A typical Raman spectrum of SWCNTs. Inset is a schematic diagram of the atomic vibrations in RBM and G-band

It is worthy noting that the peak positions and intensity of Raman spectra are sensitive to the changes of its own or external conditions. For example, the C–C bond and density of electronic states (DOS) distribution changes under various axial or radial deformations, resulting in a Raman shift or intensity change of characteristic peaks [20–22]. Consequently, resonance Raman spectroscopy can be used to analyze the deformation in SWCNTs. Similiarly, the Raman shift or intensity change of characteristic peaks is also influenced by temperature, [23] because the electron-phonon scattering is very sensitive to the temperature change. Importatnly, the Raman shift of RBM peak in large-diameter SWCNTs is more obvious than that in small-diameter ones, and even for the SWCNTs with similar diameter, the RBM peak shift is still significantly different, indicating that the temperature effect of RBM peaks is expected to be used to characterize the structure information of large-diameter SWCNTs [24]. Furthermore, the chemical doping, defects or charge transfer between SWCNTs and other materials varies the DOS distribution and Feimi level of SWCNTs, which also results in the Raman shift or intensity change of characteristic peaks [25–27]. Therefore, Raman spectroscopy has been widely used to characterize SWCNTs, which can directly or indirectly provide an abundance of information about the structure and properties of of SWCNTs, including diameter, chirality, conductive types, and so on.

1. **Diameter determination**

As an efficient and convenient technique, Raman spectroscopty has been commonly used to evaluate the diameter of SWCNTs. the RBM frequency (ω_{RBM}) has been founded to be inversely proportional to the diameter (d): [28] and the relationship between RBM frequency and diameter can be expressed by the following simple formula:

$$\omega_{\mathrm{RBM}} = \mathrm{A}/d + \mathrm{B} \tag{4.1}$$

where A is a constant related to force constants in SWCNT structures and B is a constant affected by the surrounding environment [29]. For isolated SWCNTs with different diameters or chiralities, A constant is 233–254 cm^{-1} nm and B = 0 cm^{-1} while A constant is usually 223–234 cm^{-1} nm and B is 6.5–12.5 cm^{-1} for SWCNT bundles with strong intermolecular interaction according to the previous results [30–34]. However, the linear relationship between between ω_{RBM} and d is only applicable to the SWCNTs with a diameter of 1–2 nm. For small-diameter SWCNTs (d < 1 nm): the relationship between between ω_{RBM} and d needs to be corrected according to the effects of curvature and chirality, which can be modified as [34, 35]:

$$\omega_{\mathrm{RBM}} = \mathrm{A/d} + \mathrm{B} + (\mathrm{C} + \mathrm{D}\cos^2 3\theta)/d^2 \tag{4.2}$$

where C and D corresponds to the effect of curvature, θ is chiral angle. It is worthy nting that the abovementioned constants needs to corrected according to the experimental conditions.

Expect for RBM peaks, the G-band profile is also correlated with the diameter and conductive type of SWCNTs. The G-band profile is a Lorentzian line shape for sc-SWCNTs while the G^--band has an asymmetric Breit-Wigner-Fano (BWF) line shape in the Raman spectrum of m-SWCNTs. Moreover, the G^--band of SWCNTs shows a diameter-related frequency, which can be used to roughly evaluate the diameter of SWCNTs. The relationship between G^--band, G^+-band and diameter can be expressed as following [34, 35]:

For sc-SWCNTs

$$\omega G^- = \omega G^+ - 47.7/d^2 \tag{4.3}$$

For m-SWCNTs

$$\omega G^- = \omega G^+ - 79.5/d^2 \tag{4.4}$$

2. **Identifying the conductivity type**

Since SWCNTs behavior semiconducting or metallic related with the chirality and diameter, the distinction of SWCNTs with different conductive types is very crucial before the application of SWCNTs. From the RBM frequency in Raman spectrum, the conductive type of SWCNTs with a given diametercan can also be confirmed according to the Kataura plots (Fig. 4.2a) [17]. For the mixed SWCNTs, several lasers with differnt excitation wavelengths are used to match with the interband transitions (E_{ii}) of each SWCNT, forming a resonance scattering, as shown in

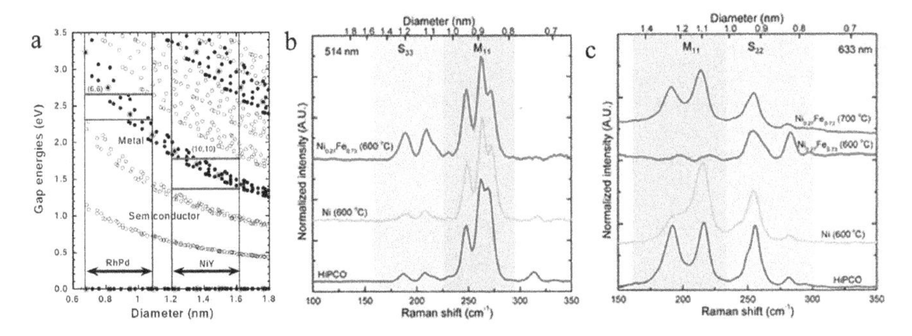

Fig. 4.2 **a** The diameter dependence of E_{ii} value for SWCNTs with diameter, that is "Kataura plots". Reproduced with permission [17]. Copyright 1999 Elsevier. **b–c** The RBM frequency of m- and sc-SWCNTs under different excitation wavelengths. Reproduced with permission [36]. Copyright 2009, American Chemical Society

Fig. 4.2b, c. When the SWCNTs is excited using 514 nm laser, the RBM frequency of m-SWCNTs can be found at 115–130 cm^{-1} and 210–285 cm^{-1} and the intermediate region corresponds to the vibrication of sc-SWCNTs. Replaced with 633 nm laser, the RBM frequency of m-SWCNTs moves to 100–110 cm^{-1} and 170 ~ 230 cm^{-1} while that of sc-SWCNTs needs to be found at other region [34, 36]. Therefore, the conductive type of mixed SWCNTs can be indentified combining RBM frequency, E_{laser} with Kataura plots [36, 37].

Furthermore, the G-band profile can also be used to distinguish sc-SWCNTs from m-SWCNTs based on the profile (symmetric or asymmetric) of G band in the Raman spectra. For all the sc-SWCNTs, both G^- and G^+ peaks show Lorentzian line shape with a full width at half-maximum (FWHM) of 6–15 cm^{-1} [38, 39]. However, the G^+ peak in m-SWCNTs still show a Lorentzian line shape while the G^- peak has a broadened BWF line shape. Moreover, the G^- peak frequency increases and the FWHM intensity decreases with increasing the diameter of m-SWCNTs [40]. Therefore, the sc- and m-SWCNTs can be easily identified from the G-band profiles under different laser wavelength excitation.

3. **The (n, m) assignment**

Importantly, the RBM mode in Raman spectra are also used to assign the (n, m) indices based on the Kataura plots which relates the diameter with E_{ii} of SWCNTs. The accurate measurement of E_{ii} value is the first problem to be solved. An energy-tunable laser is usually applied to excite the SWCNTs with a certain energy step, the RBM intensity reaches the maximum at the resonance condition with $E_{ii} = E_{Laser}$, as shown in Fig. 4.3 [24]. Using this method, the E_{ii} value, resonance window (γ) and RBM frequency of different SWCNTs can be easily determined. Another method is to simultaneously measure the Stokes and anti-Stokes RBM spectra, because the resonance windows (γ) of Stokes and anti-Stokes are very narrow (γ = 8 meV) and the resonance is asymmetric. The intensity ratios of Stokes and anti-Stokes RBM spectra are quite different for SWCNTs with similar RBM

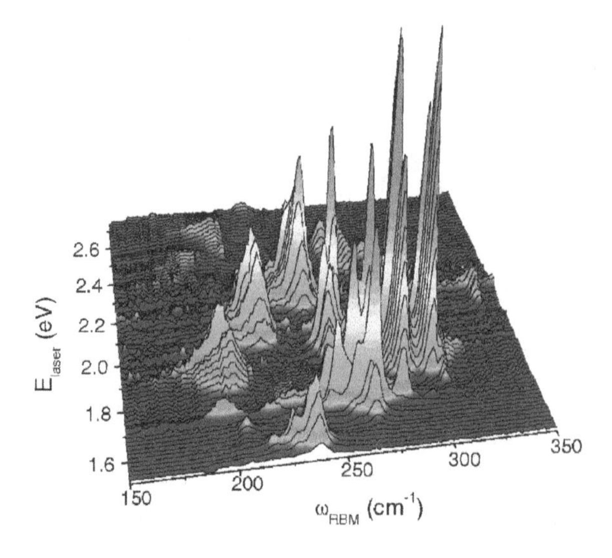

Fig. 4.3 RBM Raman mappings of SWCNTs dispersed in an SDS aqueous solution. Reproduced with permission [24]. Copyright 2004 American Physical Society

frequency, which can be used to further improve the accuracy of E_{ii} value [40, 41]. However, the (n, m) assignment based on the RBM frequency is more accurate for small-diameer SWCNTs, because the number of SWCNT at resonance conditions is very less for a given excitation energy, thereby the (n, m) indices is easily identified. For large-diameter SWCNTs, the Raman resonance occurs at the condition with E_{33} or $E_{44} = E_{Laser}$, many SWCNTs meet the resonance conditions, resulting in a big chanllenge to accurately determine (n, m) indices.

4.2.2 Graphene

Raman spectroscopy is also a powerful technique for characterizing graphene, which can be used to distinguish the few-layer graphene and its layer number from block graphite. Similar to SWCNTs, three Raman peaks appear around 1350 (D-band): 1582 (G-band) and 2700 cm cm^{-1} (2D-band) when a 514 nm laser excitation is used, respectively. In the Raman analysis, the shape, intensity and position of 2D peak are the most obvious distinction between graphene and graphite, as shown in Fig. 4.4a. Typically, the 2D peak in graphite consists of two single Lorentzian peaks ($2D_1$ and $2D_2$) with about 1/4 and 1/2 the intensity of G peak, respectively [42, 43]. While the 2D peak in monolayer graphene is a single Lorentzian peak with the intensity higher than G peak, [44–46] and the peak position of 2D band is red-shifted compared to graphite. In fact, different stackings and relative orientations can be possibly distinguished by carefully ananlyzing the shape and width of 2D peak [47, 48]. In addition, the shape and position of D peak (Fig. 4.4b) in graphene are also different from those of graphite, the graphene D peak is a single peak while two components in graphite. Moreover, both 2D and D bands are strongly dispersive with excitation energy.

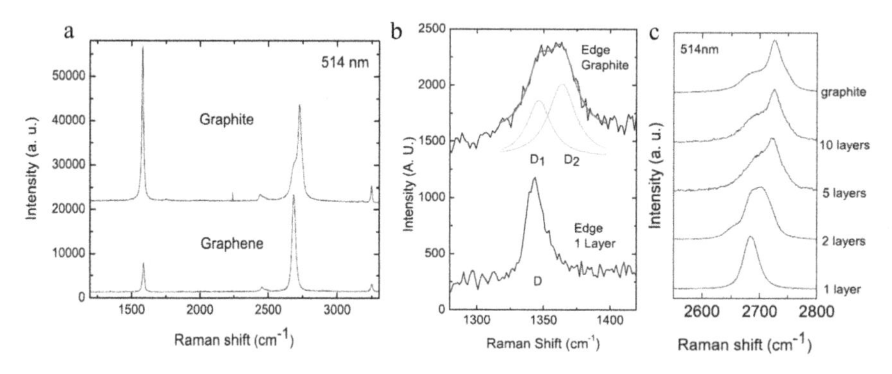

Fig. 4.4 **a** Typical Raman spectra at of graphene and graphite, and **b** their D band at 514 nm at the edge of bulk graphite and graphene. **c** The 2D band shift at 514 nm with the number of layers. Reproduced with permission [49]. Copyright 2006, American Physical Society

1. **Layer number determination**

Generally, the 2D peak shape, the G peak intensity, and the intensity ratio of G and 2D band in Raman spectra are used to determine the layer number of graphene. With the increase of number layer, the 2D peak position (Fig. 4.4c) shifts to high wavenumber (blue-shift): and both the FWHM of 2D band and the intensity ratio (I_G/I_{2D}) increase. The 2D peak shape becomes multiple Lorentzian fits, for example four Lorentzian peaks in bilayer graphene [49] and six Lorentzian peaks in tri-layer graphene [44]. Besides, the G band intensity of graphene on SiO_2/Si substrate linearly increases until approximately 10 layers ($\sim$3 nm) due to the increase of carbon atoms detected [50]. However, the the intensity of G band will decrease with the further increase of layer number due to the multiple reflection of Raman signal in graphene, as shown in Fig. 4.5. Therefore, for a few-layer graphene, the layer number can be fastly determined accoding to the Raman spectrum. In addition, the frequency ($\omega_G(n)$) of G band shifts to low wavenumber (red-shift) with the increasing layer number (n): which exhibits an almost linear dependence on 1/n, i.e., $\omega_G(n) = \omega_G(\infty) + \beta/n$, where $\beta \approx 5.5\ cm^{-1}$ is a constant and $\omega_G(\infty)$ is the wavenumber of graphite [51].

2. **Defect types and quantitative analysis of defect density**

Except for determination of layer number, Raman spectroscopy is also an effective tool to judge the defect type and defect density in graphene [52], which directly affect the fundamental physical and chemical properties of graphene. The D band of graphene corresponds to a double-resonant Raman process of the defect scattering, [43] and the intensity is closely related to the defect type and defect density. Commonly, the intensity ratio (I_D/I_G) of between the disorder-induced D-band and G-band is used to quantitatively investigate the defect type and density in graphene.

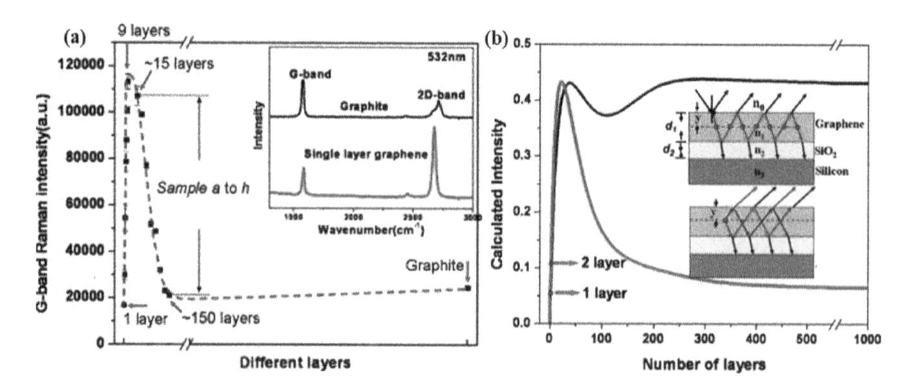

Fig. 4.5 **a** Experimental results of G-band intensity as a function of layer number. **b** The calculation results of G-band intensity as a function of layer number with and without considering the multireflection of Raman scattering light in graphene. Reproduced with permission [50]. Copyright 2002 AIP

According to Tuinstra-Koenig relationship [53], the D peak is produced only in a small region of the crystal near a defect or an edge, [43, 54, 55] which can be approximated as a point defect. For a graphene with rare defects, the intensity (I_D) of D band is proportional to the total defect number $(L_L/L_D)^2$ probed by the laser with a spot size of L_L, thus $I_D \propto (L_L/L_D)^2$, where L_D is average interdefect distance derived from scanning tunnelling microscopy. The intensity (I_G) of G band is proportional to the total spot area of laser, that is $I_G \propto L_L^2$. Consquenlty, the I_D/I_G can be expressed to be $I_D/I_G = A_{(\lambda)}/L_D^2$, where $A_{(\lambda)}$ is a coefficient related to laser wavelength. For very small L_D, $I_D/I_G = A_{(\lambda)}/L_D^2 = B_{(\lambda)}/La$, [43] where $B_{(\lambda)}$ is also a coefficient related to laser wavelength and La is the crystal size. For a known L_D in monolayer graphene, the I_D/I_G is measured to be about $145/L_D^2$ at 514 nm excitation, which is well in agreement with the above estimation [55]. According to the excitation-energy dependence of the peaks areas and intensities, the relationship between L_D and E_L^4 [I_D/I_G] can be expressed as following based on experimental results for $L_D > 10$ nm [56].

$$L_D^2\left(nm^2\right) = \frac{(4.3 \pm 1.3) \times 10^3}{E_L^4}\left(\frac{I_D}{I_G}\right)^{-1} \tag{4.5}$$

And the defect density n_D (cm^{-2}) = $10^{14}/(\pi L_D^2)$ can be obtained as following:

$$n_D\left(cm^2\right) = (7.3 \pm 2.2) \times 10^9 E_L^4\left(\frac{I_D}{I_G}\right)^{-1} \tag{4.6}$$

However, with the further increase of defect density, the Tuinstra and Koenig relation eventually fails and I_D/I_G decreases toward to 0. Meanwhile, the frequency of G band becomes dispersive with the excitation laser energy and well-defined second-order peaks disappear. For the high defect density regime with $L_D < 3$ nm, I_D decreases with respect to I_G with the decrease of L_D decreases and the I_D/I_G is proportional to $(L_D)^2$ [57]. The L_D and n_D are simply given using the following empirical formula:

$$L_D^2(nm^2) = 5.4 \times 10^{-2} E_L^4 \frac{I_D}{I_G} \tag{4.7}$$

$$n_D(cm^2) = \frac{5.9 \times 10^{14}}{E_L^4} \left(\frac{I_D}{I_G}\right)^{-1} \tag{4.8}$$

In addition, except for D band, a D′ band at aournd 1620 cm^{-1} appears in graphene with defects, which originates from an intravalley Raman scattering process. The intensity ratio ($I_D/I_{D'}$) does not depend on the defect concentration, but only on the defect types [58]. In the graphene with a low defect density, the I_D and $I_{D'}$ are proportional to the defect density, and both the intensities increase with increasing defect density. For a graphene with a certain defect density, I_D reaches a maximum and then starts to decrease, while $I_{D'}$ remains the same. Experimental results show that the $I_D/I_{D'}$ is about ~ 13 for sp^3-defects, ~ 7 for vacancy-like defects and ~ 3.5 for boundaries in graphite, respectively [52].

4.2.3 *Charge Transfer Analysis of SWCNTs and Graphene*

Except for determining structural information, Raman spectroscopy is also a powerful tool to investigate the charge transfer between carbon nanomaterials and electron donor/acceptor dopants. The concentration change of electron or hole in SWCNTs and graphene directly affects the electron–phonon coupling, thereby resulting in the change of vibration-related active bands, such as frequnency, FWHM, the relative intensity ratio, and so on [59–62].

For SWCNTs, since the G^+ band frequency is closely related to the DOS distribution and Feimi level due to the change of electron–phonon coupling, Raman spectra can be used to analyze the doping and the charge transfer in SWCNT-based heterojunctions though monitoring the Raman shift of G^+ band [18, 63–65]. Rao et al. [18] firstly investigated the doping effect of SWCNTs using Raman spectroscopty, and found that G-band shift substantially to low frequency for the electron-donor dopants (K, Rb) while to high frequency for electron-acceptor (Br_2, I_2). The charge transfer between noncovalent molecules and SWCNTs can also be monitored using the change of G-band frequency and FWHM [66]. Which has a potential to be used to design the electron-donor/acceptor molecules for the separation of sc-/m-SWCNTs, Similiarly, the sensing mechanism of SWCNT-based gas/vapor sensors will be systematically investigated based on the frequeney shift of SWCNTs exposure to different gases or vapors.

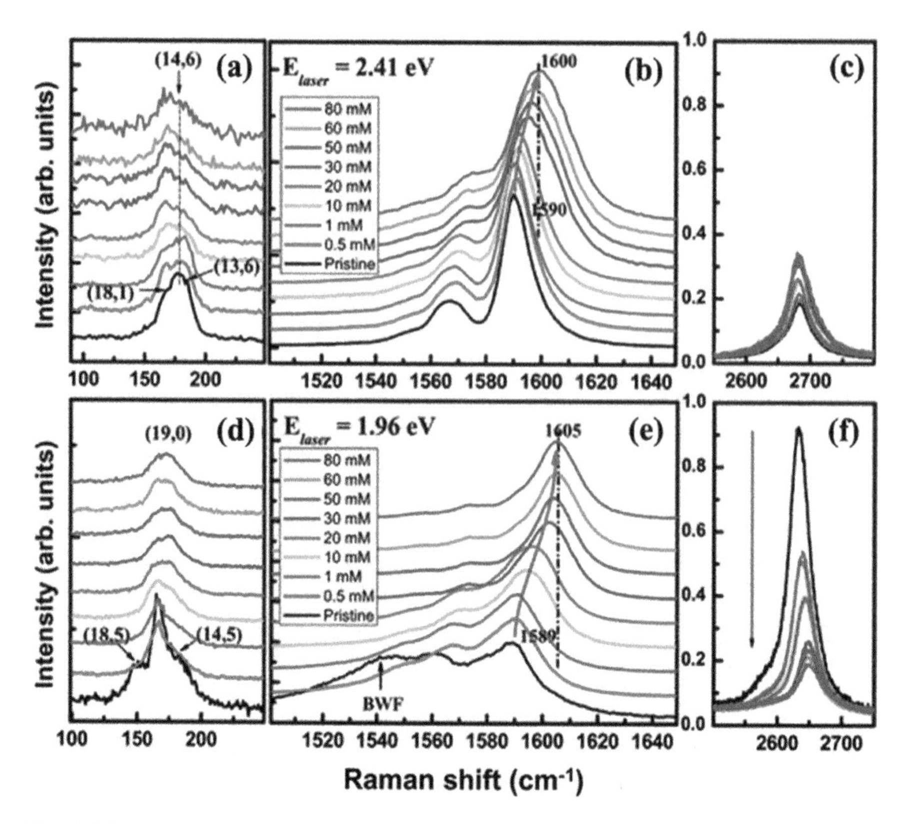

Fig. 4.6 Raman spectra of $AuCl_3$-doped SWCNTs at an excitation energy of **a–c** 2.41 eV and **d–f** 1.96 eV, respectively. Reproduced with permission [25]. Copyright 2008, American Chemical Society

In addition, Kim et al. [25] demonstrated that the G-band frequency (Fig. 4.6) was upshifted remarkably by $\sim 10\ \text{cm}^{-1}$ at a high concentration $AuCl_3$ doping, indicating a strong charge transfer from the SWCNTs to $AuCl_3$ due to the phonon stiffening effect by p-type doping. Similarly, the frequency shift of G-band of SWCNTs has been commonly used to investigate the charge transfer at the interface between SWCNTs and other materials in the heterojunctions [65, 67, 68]. For example, the G and 2D peaks of (6, 5)-enriched SWCNTs were downshifted compared to the pristine SWCNTs, indicating a static electron transfer from n-type GaAs to (6, 5)-enriched sc-SWCNTs after the formation of SWCN/GaAs heterojunctions [67]. In the (6, 5)-enriched SWCNT/graphene van der Waals (vdW) heterostructures, the 2D frequency of (6, 5)-enriched SWCNTs was upshifted while that of graphene was downshifted for SWCNT/graphene vdW heterostructures, which is attributed to the electron transfer from SWCNTs to graphene [69]. Paulus et al. [65] systematically investigated the Raman peak positions and intensities of the individual m-SWCNT and graphene in the SWCNT/graphene vdW heterostructures, and

8792 μm^{-2} electrons were transferred from SWCNT to graphene based on the analysis of the G-peak upshift of graphene. Therefore, Raman spectroscopy is a convenient powerful technique to evaluate the charge transfer dynamics at the SWCNT-based heterostructures.

Similar to SWCNTs, the G and 2D bands in graphene can also be used to estimate the charge transfer between graphene and dopants. By combining Raman spectra with the transport characteristic of graphene transistor, the G band of single or bi- layer graphene has been demonstrated to be up-shifted for both electron and hole doping [61, 70, 71]. However, for 2D band, both the frequency and relative intensity of 2D band are quite sensitive to doping, and the frequency shows a hole-doping related upshifts and electron-doping related down-shifts [61]. Voggu et al. [62] demonstrated that the G-band softened progressively with increasing the electron-donor concentration (tetrathiafulvalene, TTF) while the band stiffened with increasing the electron-acceptor concentration (tetracyanoethylene, TCNE). Meanwhile, the FWHM of G band increased due to the interaction between graphene and the molecules. Importantly, the intensity of 2D band and the I_{2D}/I_G decreased with increasing the concentration of TTF or TCNE molecules. Dong et al. [72] suggested that the doping level (or Fermi energy) of monolayer graphene could be modulated using aromatic molecules. The hole doping resulted in the upshifts of both G and 2D band, whereas the electron doping upshifted the 2D band but downshifted the G band. More importantly, the electronic effect has been considered to dominate the stiffening or softening of the G band, rather than the dynamic effect observed in electrical doping. It is wothy noting that the charge transfer induced by molecule adsorption seems to be in contrast to that by gate-induced or chemical doping, and much more efforts are still required.

4.3 Electron Microscopy

4.3.1 SWCNTs

As we know, the tubular structure of CNTs was firstly observed from an transmission electron microscopy (TEM) image in 1991, [1] since then TEM has been commonly used to determine the wall number and diameter of CNTs. With the further understanding of SWCNT characterizations, the detailed arrangement of carbon atoms in a SWCNT can be clearly visualized using high-resolution TEM (HRTEM) at an atomic-level resolution, from which the chiral information and structural defects of individual SWCNTs have been obtained [73–76]. Meanwhile, the fast Fourier transformation (FFT) patterns derived from their TEM images can also be used to further determine the chiral indices of the SWCNTs, [76] which have characteristic bright spots forming hexagons. By combining TEM image with FFT pattern (Fig. 4.7a): the chiral indices of SWCNTs were succussfully determined to be (10, 10) and (11, 8): as shown in Fig. 4.7b, c. Moreover, one can

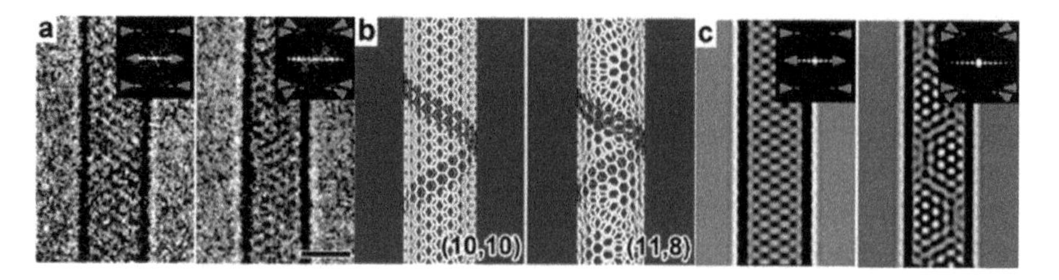

Fig. 4.7 **a** TEM images of (10, 10) and (11, 8) SWCNTs SWCNTs and their FFT images. **b** The corresponding structure diagrams. **c** Simulated TEM images and their FFT images. Reproduced with permission [76]. Copyright 2008, American Chemical Society

distinguish SWCNTs with identical structures but different helical angles, thereby both diameter and chirality distribution of SWCNTs in the sample can be statistically analyzed using this technique. However, the TEM sample preparation and atomic-level imaging are extremely demanding process, limiting the the further improvement of characterization efficiency.

Comparatively, nanobeam electron diffraction (ED) is a more convenient technique in determining the atomic structure of SWCNTs, [77–80] which was first used to identify the helical feature of SWCNTs in 1991. And then, a complete kinematical ED theory based on an individual SWCNT was formulated by Qin in 1994 [77], in which the ED pattern of a SWCNT can be simply considered as diffractions from two parallel graphene sheets, corresponding to the top and bottom wall of SWCNTs. Clearly, if a graphene sheet is rotated along the normal direction of the plane relative to another sheet, the ED pattern will remain the same due to same rotation angle. The radial curvature of SWCNTs makes the diffraction spots appear distribted in the radial direction, forming a series of discrete lines (layer lines) [77]: as shown in Fig. 4.8. The (n, m) assignment and chiral angle of SWCNTs can be determined from the intensity distribution and the ratio of peak spacing on each principal layer line in the ED patterns. The radial intensity distribution of each layer line is described by the square of a Bessel function ($J_N(\pi dX)$): that is, the intensities of three principal layer lines (l_1, l_2, and l_3) are proportional to $|J_{n+m}(\pi dX)|^2$, $|J_n(\pi dX)|^2$, and $|J_m(\pi dX)|^2$, respectively. The order N (n, m or n + m) can be rapidly and accurately determined by comparing the theoretical value with the measured peak spacing ratio (P_2/P_1) of its second and first order peaks, respectively. Thereby, the chiral indices (n, m) of SWCNTs are directly assigned. For a zigzag (n, 0) SWCNT, the principal layer lines l_1 and l_2 overlap with intensities proportional to $|J_n(\pi dX)|^2$, while the intensity distribution of layer line l_3 is proportional to $|J_0(\pi dX)|^2$. For an armchair (n, n) SWCNT, the principal layer lines l_2 and l_3 overlap with each other, and its intensity distribution is proportional to $|J_n(\pi dX)|^2$, while the layer line l_1 overlaps with the equatorial line. However, using this method, the relative orientation between SWCNT and incident electron beam has an extremely requirement, any deviation will causes a significant change in the intensity distribution of peak positions, thereby resulting in a failure of the Bessel function assignment [40].

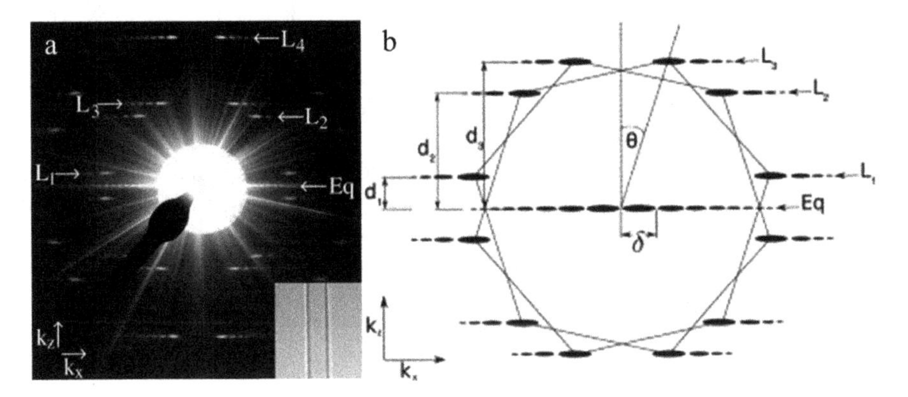

Fig. 4.8 **a** Selected area diffraction pattern from an individual SWCNT and **b** schematic depiction of the diffraction pattern obtained from a SWCNT. Reproduced with permission [80]. Copyright 2011, Elsevier

Since the positions of each layer line in the axial direction are unaffected by the radial curvature of SWCNTs, the axial positions of the layer lines have be demonstrated to determine chiral structure of SWCNTs [81]. The chiral angle (θ) of helical SWCNTs can be obtained as following [80, 81]:

$$\theta = \tan^{-1}\left(\frac{d_2 - d_1}{\sqrt{3}d_3}\right) = \tan^{-1}\left(\frac{2d_2 - d_3}{\sqrt{3}d_3}\right) \tag{4.9}$$

where d_i represents the axial distance of the l_i layer line to the equatorial line. Importantly, these axial distances are closely related to the chiral indices, and the relationshop between d_i and the chiral indices can be expressed using the following foumula [79, 80, 82].

$$\frac{m}{n} = \frac{d_2 - d_1}{d_2 + 2d_1} = \frac{d_3 - 2d_1}{d_3 + d_1} = \frac{2d_2 - d_3}{2d_3 - d_2} \tag{4.10}$$

The ratio m/n is independent of the camera length and the relative orientation between the nanotube and the incident electron beam. Even though the signal/noise ratio is low, the ratio can still be conveniently obtained [79]. Therefore, when the layer line spacings are accuately measured from ED patterns, the chiral indices (n, m) can also be easily determined with a very high accuracy. Jiang et al. [83] further proposed an absolutely calibration-free approach to determine the chiral indices of SWCNTs based on a new concept of "intrinsic layer-line spacing", in which only the layer line spacings d_i and the interval δ between the zeros along the equatorial line are involved in the measurement. Using this method, the tilt angle of SWCNTs with respect to the electron beam can be simultaneously evaluated, and the tilt effect is automatically compensated.

4.3.2 Graphene

The microstructure of graphene has also been widely characterized using TEM, and the crystal lattice and layer number can be clearly observed from a HRTEM image. Especailly, the atomic level defects and fine edge structures can be accurately resolved without destroying the stability of graphene using spherical aberration-corrected HRTEM (AC-HRTEM): which will give direct evidence to understand the electrical, chemical, magnetic and mechanical properties of graphene. As shown in Fig. 4.9, Meyer et al. [84] firstly observed the formation and real-time dynamics of Stone–Wales defects using AC-HRTEM with an electron acceleration voltage of 80 kV and a resolution of 0.1 nm, and it has been found that the multiple pentagon-heptagon (5–7) defects relax to the original unperturbed lattice, which has been demonstrated to play a key role in the formation and transformation of sp^2 bonded carbon nanostructures. By combining with the

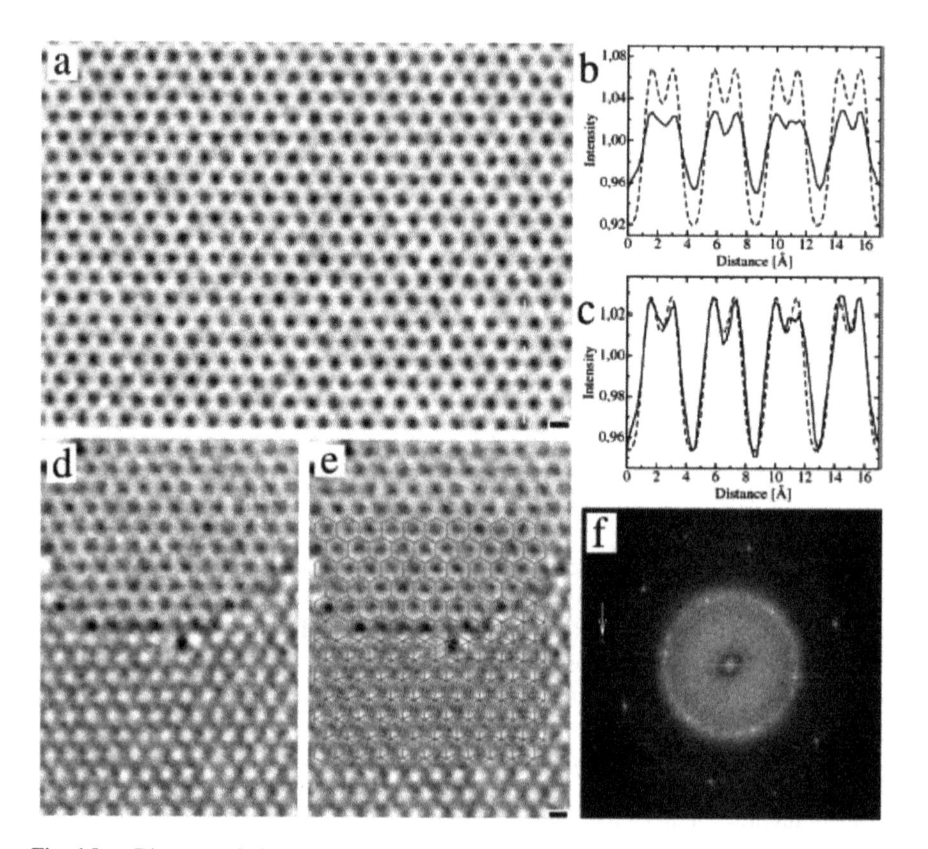

Fig. 4.9 **a** Direct atomic image of a monolayer graphene. **b–c** Contrast profile along the dotted line in panel a (solid) along with a simulated profile (dashed). **d–e** Step from a monolayer (upper part) to a bilayer (lower part). **f** Numerical diffractogram calculated from an image of the bilayer region. Reproduced with permission [84]. Copyright 2008, American Chemical Society

analysis and filtering of 2D FFT, the as-generated reconstructed images can be used to elucidate the stacking configuration of the few-layer graphene. Furthermore, the Moiré pattern in the AC-HRTEM images revealed two discernible types of graphene boundaries in the few-layer graphene system, [85] (1) graphene sheets from one domain grown over the top of a neighboring domain, and (2) graphene domains interconnected by directly atomic bonding two graphene sheets with different lattice directions. Warner et al. [86] further resolved the atomic structure of ABC rhombohedral trilayer graphene, and indentified the relative stacking orientation between multiple graphene layered structures.

The selected-area electron diffraction (SAED) is also commonly used to characterize the crystal structure of graphene and accurately indentify monolayer grapheme [87]. For monomorph graphene with mono- or multi-layers, the six-fold symmetry of graphene results in one set of six-fold-symmetric spots, as shown in Fig. 4.10. Intenstingly, the relative intensity of the {1100} and the {2110} diffraction spots can be used to distinguish the monolayer graphene from multi-layer graphene, and the intensity of the {1100} diffraction spots higher than the {2110} is a unique characteristic for monolayer graphene. Moreover, The intensity ratios ($I_{\{1100\}}/I_{\{2110\}}$) for monolayer and bilayer graphene are 1.4 and 0.4, respectively, [87] which can be used to roughly evaluate the yield of monolayer graphene. In addition, the electron diffraction is sensitive to the structural changes on surface of graphene, such as intrinsic or surface-adsorption induced ripples. Therefore, the SAED pattern of graphene can also be used to analyze the micro-structural changes on the surface and edge of graphene. For polycrystalline graphene, the SAED patterns constain many circularly distributed diffraction spots, corresponding to different in-plane lattice orientations. Importantly, by combining electron diffraction with diffraction-filtered imaging, the location, orientation and shape of several hundred grain domains in graphene can be fast determined without imaging billions of atoms in each domain, [88] as shown in Fig. 4.11. The method provides a powerful tool to characterize graphene grains and grain boundaries on all relevant length scales, which is also crucial for exploring synthesis strategies and optimizing physical properties of graphene.

4.4 UV-vis-NIR Absorption Spectroscopy

Expect for Raman spectra, the interband electronic transitions can also be characterized from UV-vis-NIR absorption spectra. Different from conventional bulk carbon materials, the absorption spectra of SWCNTs shows a sharp profile due to unique quasi-1D structural confinement effect, and the characteristic absorption peaks induced by interband E_{ii} transitions can be clearly observed for a given isolated SWCNT. The E_{ii} transition-dependent UV-vis-NIR absorption spectrum has been widely used to identify the diameter and chirality distribution of SWCNTs. Normally, three interband optical transitions (S_{11}, S_{22} and S_{33}) for sc-SWCNTs and the first interband optical transition (M_{11}) for m-SWCNTs can be

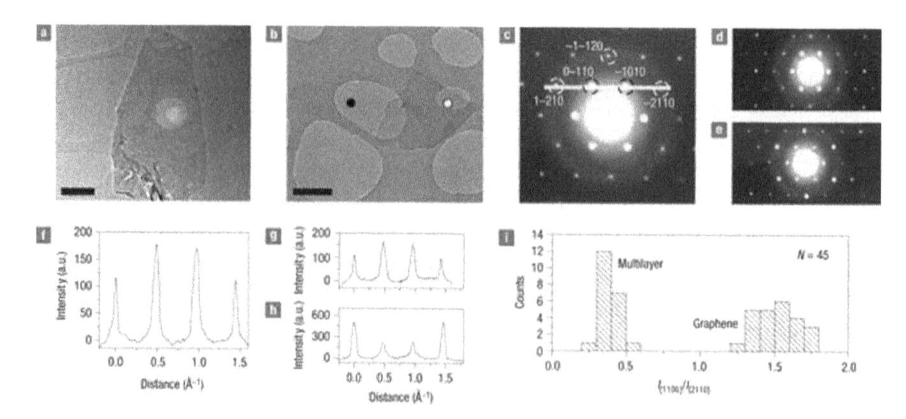

Fig. 4.10 HRTEM images of **a** monolayer and **b** bilayer graphene. SAED patterns of **c** monolayer and **d–e** bilayer graphene taken from the positions of the black and white spots, respectively. **f–h** The corresponding diffracted intensity taken along the 1–210 to -2110 axis for the patterns in **c–e**, respectively. **i** Histogram of the intensity ratios of the {1100} and {2110} diffraction peaks for all the diffraction patterns collected. Reproduced with permission [87]. Copyright 2007, Springer Nature

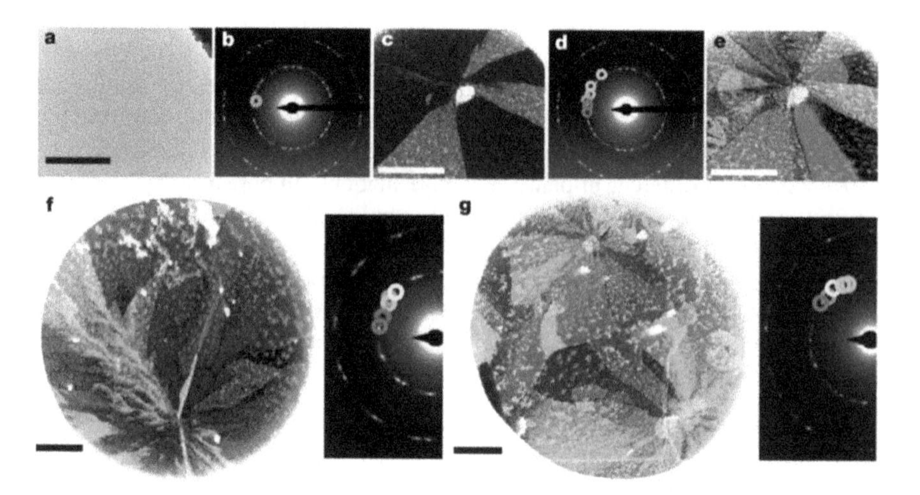

Fig. 4.11 **a–e** Grain imaging process of polycrystalline graphene based on bright-field TEM images and SAED. **f–g** Grain images of polycrystalline graphene. Reproduced with permission [88]. Copyright 2011, Springer Nature

seen from the UV-vis-NIR absorption spectra. For most of sc-SWCNTs, the S_{11} optical transition results in a strong absorption in the NIR region ranging from 800 to 1800 nm, and the light absorption induced by S_{22} optical transition occurs in the vis-NIR region ranging from 550 to 1100 nm. For m-SWCNTs, the absorption peaks are usually observed in the visible light region ranging from 400 to 750 nm.

Importantly, these optical transitions provide qualitative and semi-quantitative information about the diameter, relative purity, and the ratio of semiconducting SWCNTs to metallic ones.

To accurately indentify the diameter distribution and (n, m) indices in mixed SWCNTs, the SWCNTs powders are usually dispersed in aqueous solution containing surfactant under ultrasonic condition, and then ultrahigh-speed centrifugation is applied to remove the SWCNT bundles and residual catalyst particles, leaving monodisperse SWCNTs in the supernatant. The diameter and diameter distribution of mixed SWCNTs can be obtained from the position of each absorption peak, which can be approximatively obtained by the following equations: $E_{11} = 0.96/d$ and $E_{22} = 1.70/d$ for sc-SWCNTs, while $E_{11} = 2.60/d$ for m-SWCNTs. However, other factors have to be considered comprehensively before the precise determination of SWCNT diameter, such as type and concentration of surfactant, solvent, pH value and so on. Combining tight-binding band-structure calculations with a chirality- and diameter-dependent nearest-neighbor hopping integral, Hagen et al. [8] demonstrated the assignment of (n, m) indices of 14 SWCNTs from UV-vis-NIR obsorption spectra of individual SWCNTs, supporting a quantitative analysis of absorption spectrum. Meanwhile, the abundances of SWCNTs with different (n, m) indices can also be estimated from the absorption spectrum through a quantitative analysis, For example, Yang et al. [3] reported a chirality-specific growth sc-SWCNTs on a solid bimetallic alloy catalyst, and the content of (12, 6) SWCNTs is 92.5% according to the obsorption spectrum ranging from 450 to 800 nm.

In addition, UV-vis-NIR absorption spectrscopy is also a facile powerful approach to quantitatively analyze the relative purity of SWCNTs in samples, as shown in Fig. 4.12. The absorption peak of S_{22} transition is usually chosen to be analyzed rather than S_{11} transition, because it is less affected by doping or other factors and also matches the transmission window for DMF [89]. The relative purity can be evaluated using a simple ratio of 0.141A(S)/A(T) [90]: where A(S) is the area of the S_{22} spectral band after linear baseline correction and A(T) is the total area under the S_{22} band curve. Similarly, the ratio of semiconducting nanotubes to metallic ones in bulk SWCNTs can be easily evaluated without pure sc-SWCNTs as standard reference [6, 91]. Experimentaly, one needs to draw a straight line joining the minima on each side of the S_{11} and M_{11} peaks and calculate the peak areas between the spectrum curve and the straight line. The relative content (R_{sc}) of sc-SWCNTs can be calculated according to the integral peak areas of the interband S_{11} and M_{11} transitions, which is given by:

$$R_{sc} = \frac{A(S_{11})}{A(S_{11}) + A(M_{11})} \tag{4.11}$$

where $A(S_{11})$ and $A(M_{11})$ are the integral peak areas of the interband S_{11} and M_{11} transitions, respectively. Although the content does not represent the precise ratio of sc-SWCNTs due to complex absorption characteristics (e.g. extinction coefficients) and diameter distribution of sc-/m-SWCNTs, Semi-quantitative comparison still provides the ratio of sc-SWCNTs in different samples within similar SWCNTs.

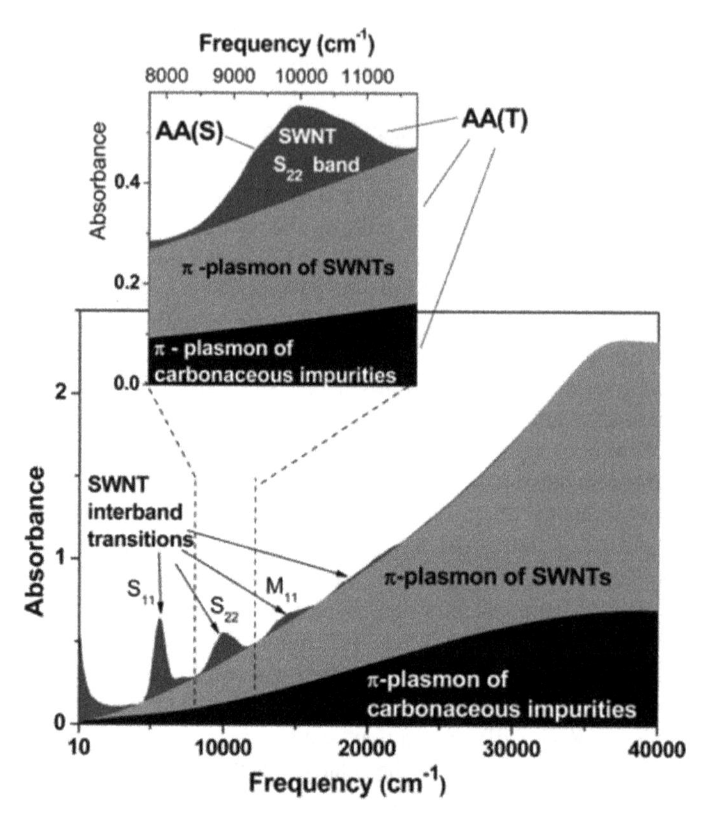

Fig. 4.12 Schematic illustration of the electronic spectrum of a typical SWCNT sample produced by the electric arc method. Reproduced with permission [90]. Copyright 2005, American Chemical Society

4.5 Photoluminescence Spectroscopy

Except for Raman and absorption spectroscopy, photoluminescence (PL) spectroscopy as a complement has also been commonly used to assign the (n, m) indices of sc-SWCNTs and to evaluate the SWCNT abundance, especially for sc-SWCNTs with similar E_{22} or E_{11} energies (Fig. 4.13) [7: 12899]. Berifly, an electron in an SWCNT is excited from v_2 state to c_2 state via interband E_{22} transition, generating an exciton (electron-hole pair). Subsequently, the electron at c_2 state and hole at c_2 state rapidly relax to c_1 and v_1 states in the femtosecond scale, respectively. Finally, the electron-hole recombination via interband E_{11} transition ($c_1 - v_1$): emitting a photon with the E_{11} energy. In most cases, the emission wavelengths of sc-SWCNTs are in the NIR region ranging from 800 to 1800 nm, while the PL process cannot be generated in m-SWCNTs without bandgap. Thus, PL spectra are only used to analyze the chirality or abundance of sc-SWCNTs, and

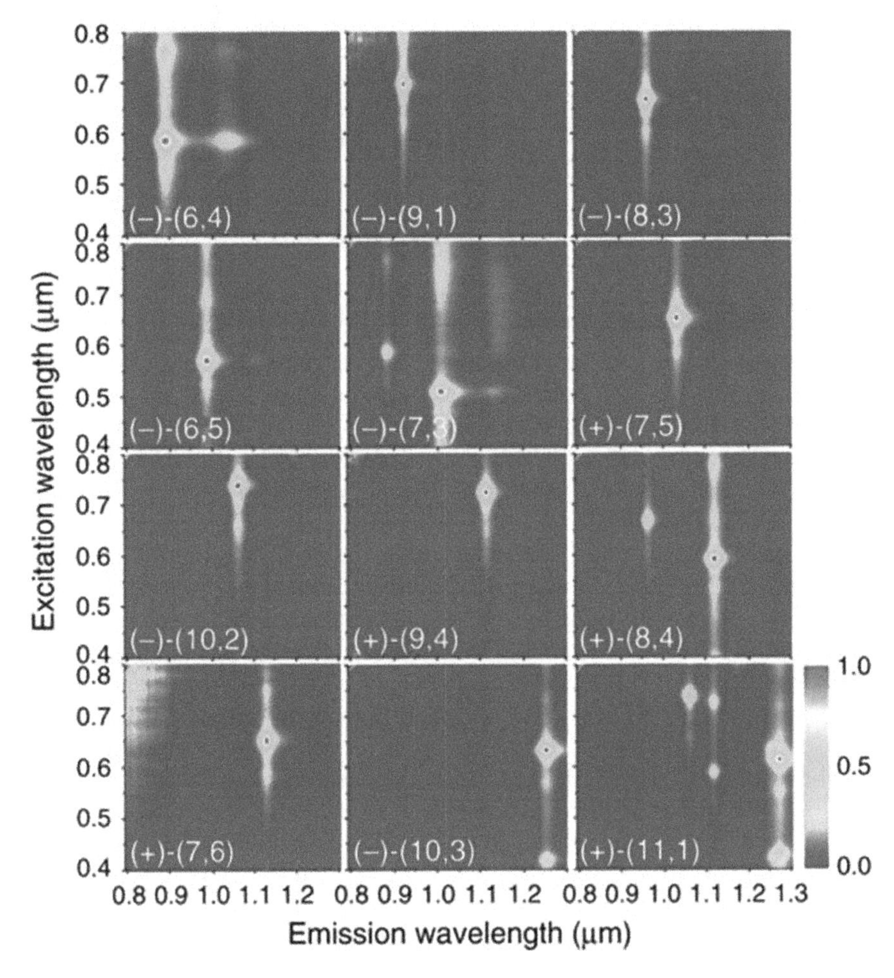

Fig. 4.13 PL contour maps of the 12 sorted (n, m) sc-SWCNTs. Reproduced with permission [32]. Copyright 2016, Nature

isolated SWCNTs are required in an aqueous surfactant suspension due to the existence of fluorescence quenching in SWCNT bundles.

As we know, the PL intensity is related to the electronic structures and content of sc-SWCNTs, wavelength and intensity of excitation light, as well as the sensitivity of NIR photodetector. For a SWCNT with specific chirality, the intensity of fluorescence with the E_{11} energy varies remarkably with changing the excitation wavelength, and the fluorescence intensity reaches the maximum when the excitation photon energy matches the E_{22} energy of a SWCNT. Consequently, a contour plot (or PLE mapping) can be produced according to the relationship between the PL intensity and excitation and emission wavelengths, in which the discrete PL peaks arise from the (n, m) indices of specific sc-SWCNTs. Experimentally, the diameter and chiral angle of a sc-SWCNTs can be obtained from the following formula [32]:

$$E_{11} = \frac{hc}{\lambda_{11}} = hcv_{11} = hc\left(\frac{10^7}{157.5 + 1066.9d} + \frac{A_1 \cos 3\theta}{d^2}\right) \tag{4.12}$$

$$E_{22} = \frac{hc}{\lambda_{22}} = hcv_{22} = hc\left(\frac{10^7}{157.5 + 1066.9d} + \frac{A_2 \cos 3\theta}{d^2}\right) \tag{4.13}$$

where h and c are Planck's constant and light speed, respectively. Photon wavelength, and photon frequency in cm^{-1}. λ_{22} (or v_{22}) and λ_{11} (or v_{11}) are the excitation wavelength (or frequency) and emission wavelength (or frequency). Both A_1 and A_2 are the constant depended on the remainder of (n − m)/3, A_1 and A_2 are −710 and 1375 cm^{-1} for the remainder = 1, respectively. While the A_1 and A_2 values are 369 and −1475 cm^{-1} for the remainder = 2, respectively. Moreover, the sc-SWCNT family with a remainder of 2 display higher PL emission intensities than those with a remainder of 1 [92]. Furthermore, Luo et al. [93] developed a methodology to evaluate the (n, m) abundance of sc-SWCNTs from a PL excitation map, in which the dot size of the PL peak is proportional to the PL intensity of each sc-SWCNT. Using a single-particle electron–phonon interaction model, the experimentally PL intensities sc-SWCNTs could be corrected and the abundance of a specific (n, m) SWCNT would be calculated, obtaining the diameter and chirality distributions of sc-SWCNTs. Therefore, the PLE mapping has been widely used to precise (n, m) assignment. Howcver, it is mainly used to characterize the small-diameter sc-SWCNTs due to large excitation-energy dependent fluorescence. The (n, m) assignment of large-diameter sc-SWCNTs is still a big challenge. Actually, the PL peak and intensity of sc-SWCNTs dispersed in solutions are easily influenced by the solvent, surfactant, pH value in solution, temperature and so on, [94–98] the accurate (n, m) assignment of sc-SWCNTs usually needs the assistance of absorption and Raman spectroscopy. Furthermore, PL analysis of the sc-SWCNTs on substrates is also very important because the fluorescence excitation and quenching can be used to investigate the charge transfer dynamics at the interface between sc-SWCNTs and other materials. However, the PL intensity of sc-SWCNTs on substrates is usually very weak due to the interactions between SWCNTs and substrate or defect states [40]. Although the PL emission can be recovered by self-assembling monolayers on substrate, [99] new effort to further improve the PL efficiency of sc-SWCNTs is still needed in the future.

4.6 Other Characterization Techniques

Atomic force microscopy (AFM) is usually used to characterize the micro morphologies of SWCNTs and graphene on the smooth substrate, such as diameter/height, length, density or orientation. However, it can not determine the detailed structural information of a SWCNT and graphene. It is well known that scanning tunneling microscopy (STM) can be used to characterize the chiral structures of SWCNTs due to its atomic-level resolution. Ge et al. [100, 101] first demonstrated a STM characterization of SWCNTs on highly oriented pyrolytic graphite, visualizing a precise location of carbon atoms and determining the helical structure of SWCNTs. The atomic arrangement, chiral indices and the corresponding electronic properties have been further inverstigaed using this technique [102, 103]. The chiral angle of SWCNTs can be easily determined by measuring the zigzag tube axis direction relative to the nanotube axis. Combining this chiral angle with measured diameter, one can assign the (n, m) indices of a SWCNT. Importantly, the electronic properties of isolated SWCNT can be directly analyzed by measuring the tunneling current at different bias voltages between nanotube and the STM tip. The curve of normalized conductance versus bias voltage can be used to measure the main features in the local density of electronic states (LDOS) for m- and sc-SWCNT, such as the conduction and valence band edges, bandgap, etc. Futhermore, the subtle change of atomic structure in SWCNT can be distinguished from an atomic-resolution STM image, which is helpful to inverstigate the intramolecular junction of SWCNTs [104]. The detailed atomic assignment of the junction can not only be clearly characterized, but the electrical properties of the intramolecular junction can also be quantitatively measured, providing an insight into SWCNT heterojunctions for nanoelectronics. Howcver, the STM characterization requires ultra-high vacuum conditions and the flat substrate at atomic level, resulting in a very low efficiency and mun inconvenience in use.

References

1. Iijima S (1991). Helical microtubules of graphitic carbon. Nature, 354(6348): 56–58.
2. Iijima S, Ichihashi T (1993). Single-shell carbon nanotubes of 1-nm diameter. Nature, 363 (6430): 603–605.
3. Yang F, Wang X, Zhang D, Yang J, Luo D, Xu Z, Wei J, Wang J, Xu Z, Peng F, Li X, Li R, Li Y, Li M, Bai X, Ding F, Li Y (2014): Chirality-specific growth of single-walled carbon nanotubes on solid alloy catalysts. Nature, 510(7506): 522–524.
4. Li X, Tu X, Zaric S, Welsher K, Seo WS, Zhao W, Dai H (2007). Selective synthesis combined with chemical separation of single-walled carbon nanotubes for chirality selection. J Am Chem Soc, 129(51): 15770–15771.
5. Liu H, Nishide D, Tanaka T, Kataura H (2011). Larye-scale single-chirality separution of single-wall carbon nanotubes by simple gel chromatography. Nat Commun, (2): 309.
6. Miyata Y, Yanagi K, Maniwa Y, Kataura H (2008). Optical evaluation of the metal-to-semiconductor ratio of single-wall carbon nanotubes. J Phys Chem C, 112(34): 13187–13191.
7. Itkis M E, Perea DE, Niyogi S, Rickard SM, Hamon MA, Hu H, Zhao B, Haddon RC (2003). Purity evaluation of as-prepared single-walled carbon nanotube soot by use of solution-phase near-IR spectroscopy. Nano Lett, 3(3): 309–314.
8. Hagen A, Hertel T (2003). Quantitative analysis of optical spectra from individual single-wall carbon nanotubes. Nano Lett, 3(3): 383–388.

9. Bao W, Jing L, Velasco J, Lee Y, Liu G, Tran D, Standley B., Aykol M., Cronin SB, Smirnov D, Koshino M, McCann E, Bockrath M, Lau CN (2011). Stacking-dependent band gap and quantum transport in trilayer graphene. Nat Phys, 7(12): 948–952.
10. Cao Y, Fatemi V, Fang S, Watanabe K, Taniguchi T, Kaxiras E, Jarillo-Herrero P (2018). Unconventional superconductivity in magic-angle graphene superlattices. Nature, 556 (7699): 43–50.
11. Park JM, Cao Y, Watanabe K, Taniguchi T, Jarillo-Herrero P (2021). Tunable strongly coupled superconductivity in magic-angle twisted trilayer graphene. Nature, 590(7845): 249–255.
12. Kim K, Lee Z, Regan W, Kisielowski C, Crommie MF, Zettl A (2011). Grain boundary mapping in polycrystalline graphene. ACS Nano, 5(3): 2142–2146.
13. Rasool HI, Ophus C, Klug WS, Zettl A, Gimzewski JK (2013). Measurement of the intrinsic strength of crystalline and polycrystalline graphene. Nat Commun, 4(1): 1–7.
14. Dresselhaus MS, Dresselhaus G, Saito R, Jorio A (2005). Raman spectroscopy of carbon nanotubes. Phys Rep, 409(2): 47–99.
15. Eklund PC, Holden JM, Jishi RA (1995). Vibrational modes of carbon nanotubes; spectroscopy and theory. Carbon, 33(7): 959–972.
16. Rao AM, Richter E, Bandow S, Chase B, Eklund PC, Williams KA, Fang S, Subbaswamy KR, Menon M, Thess A, Smalley RE, Dresselhaus G, Dresselhaus MS (1997). Diameter-selective Raman scattering from vibrational modes in carbon nanotubes. Science, 275(5297): 187–191.
17. Kataura H, Kumazawa Y, Maniwa Y, Umezu I, Suzuki S, Ohtsuka Y, Achiba Y (1999). Optical properties of single-wall carbon nanotubes. Synthetic Metals, 103(1–3): 2555–2558.
18. Rao AM, Eklund PC, Bandow S, Thess A, Smalley RE (1997). Evidence for charge transfer in doped carbon nanotube bundles from Raman scattering. Nature, 388(6639): 257–259.
19. Zhang D, Yang J, Li M, Li Y (2016). (n, m) assignments of metallic single-walled carbon nanotubes by Raman spectroscopy: The importance of electronic Raman scattering. ACS Nano, 10(12): 10789–10797.
20. Yang L, Han J (2000). Electronic structure of deformed carbon nanotubes. Phys Rev Lett, 85 (1): 154.
21. Lucas M, Young RJ (2004). Effect of uniaxial strain deformation upon the Raman radial breathing modes of single-wall carbon nanotubes in composites. Phys Rev B, 69(8): 085405.
22. Cronin SB, Swan AK, Ünlü MS, Goldberg BB, Dresselhaus MS, Tinkham M (2004). Measuring the uniaxial strain of individual single-wall carbon nanotubes: resonance Raman spectra of atomic-force-microscope modified single-wall nanotubes. Phys Rev Lett, 93(16): 167401.
23. Raravikar NR, Keblinski P, Rao AM, Dresselhaus MS, Schadler LS, Ajayan PM (2002). Temperature dependence of radial breathing mode Raman frequency of single-walled carbon nanotubes. Phys Rev B, 66(23): 235424.
24. Fantini C, Jorio A, Souza M, Strano M, Dresselhaus M, Pimenta M (2004). Optical transition energies for carbon nanotubes from resonant Raman spectroscopy: Environment and temperature effects. Phys Rev Lett, 93: 147406.
25. Kim KK, Bae JJ, Park HK, Kim SM, Geng HZ, Park KA, Shin HJ, Yoon SM, Benayad A, Choi JY, Lee YH (2008). Fermi level engineering of single-walled carbon nanotubes by $AuCl_3$ doping. J Am Chem Soc, 130(38): 12757–12761.
26. Chiu PW, Duesberg GS, Dettlaff-Weglikowska U, Roth S (2002). Interconnection of carbon nanotubes by chemical functionalization. Appl Phys Lett, 80(20): 3811–3813.
27. Gupta S, Hughes M, Windle AH, Robertson J (2004). Charge transfer in carbon nanotube actuators investigated using in situ Raman spectroscopy. J Appl Phys, 95(4): 2038–2048.
28. Jishi R, Venkataraman L, Dresselhaus M, Dresselhaus G (1993). Phonon modes in carbon nanotubules. Chem Phys Lett, 209: 77–82.
29. Liu K, Wang W, Wu M, Xiao F, Hong X, Aloni S, Bai X, Wang E, Wang F (2011). Intrinsic radial breathing oscillation in suspended single-walled carbon nanotubes. Phys Rev B, 83: 113404.

30. Cheng HM, Li F, Sun X, Brown SDM, Pimenta MA, Marucci A, Dresselhaus G, Dresselhaus MS (1998). Bulk morphology and diameter distribution of single-walled carbon nanotubes synthesized by catalytic decomposition of hydrocarbons. Chem Phys Lett, 289(5–6): 602–610.
31. Alvarez L, Righi A, Guillard T, Rols S, Anglaret E, Laplaze D, Sauvajol JL (2000). Resonant Raman study of the structure and electronic properties of single-wall carbon nanotubes. Chem Phys Lett, 316(3–4): 186–190.
32. Bachilo SM, Strano MS, Kittrell C, Hauge RH, Smalley RE, Weisman RB (2002). Structure-assigned optical spectra of single-walled carbon nanotubes. Science, 298(5602): 2361–2366.
33. Milnera M, Kürti J, Hulman M, Kuzmany H (2000). Periodic resonance excitation and intertube interaction from quasicontinuous distributed helicities in single-wall carbon nanotubes. Phys Rev Lett, 84(6): 1324.
34. Zhang Y, Zhang J (2012). Application of resonance Raman spectroscopy in the characterization of single-walled carbon nanotubes. Acta Chim Sinica, 70: 2293–2305.
35. Dresselhaus MS, Dresselhaus G, Jorio A (2004). Unusual properties and structure of carbon nanotubes. Annu Rev Mater Res, 34: 247–248.
36. Chiang WH, Sakr M, Gao XP, Sankaran RM (2009). Nanoengineering Ni_xFe_{1-x} catalysts for gas-phase, selective synthesis of semiconducting single-walled carbon nanotubes. ACS Nano, 3(12): 4023–4032.
37. Weisman RB, Bachilo SM (2003). Dependence of optical transition energies on structure for single-walled carbon nanotubes in aqueous suspension: an empirical Kataura plot. Nano Lett, 3(9): 1235–1238.
38. Dresselhaus MS, Dresselhaus G, Jorio A, Souza Filho AG, Saito R (2002). Raman spectroscopy on isolated single wall carbon nanotubes. Carbon, 40(12): 2043–2061.
39. Shang NG, Silva SRP, Jiang X, Papakonstantinou P (2011). Directly observable G band splitting in Raman spectra from individual tubular graphite cones. Carbon, 49(9): 3048–3054.
40. He M, Zhang S, Zhang J (2020). Horizontal single-walled carbon nanotube arrays: controlled synthesis, characterizations, and applications. Chem Rev, 120: 12592–12684.
41. Dresselhaus MS, Dresselhaus G, Jorio A, Souza Filho AG, Pimenta M A, Saito R (2002). Single nanotube Raman spectroscopy. Acc Chem Res, 35(12): 1070–1078.
42. Nemanich RJ, Solin SA (1979). First- and second-order Raman scattering from finite-size crystals of graphite. Phys Rev B, 20: 392–401.
43. Ferrari AC, Basko DM (2013). Raman spectroscopy as a versatile tool for studying the properties of graphene. Nat Nanotechnol, 8(4): 235–246.
44. Malard LM, Pimenta MA, Dresselhaus G, Dresselhaus MS (2009). Raman spectroscopy in graphene. Phys Rep, 473(5–6): 51–87.
45. Park JS, Reina A, Saito R, Kong J, Dresselhaus G, Dresselhaus MS (2009). G′ band Raman spectra of single, double and triple layer graphene. Carbon, 47: 1303–1310.
46. Graf D, Molitor F, Ensslin K, Stampfer C, Jungen A, Hierold C, Wirtz L (2007). Spatially resolved Raman spectroscopy of single- and few-layer graphene. Nano Lett, 7(2): 238–242.
47. Lui CH, Li Z, Chen Z, Klimov PV, Brus LE, Heinz TF (2011). Imaging stacking order in few-layer graphene. Nano Lett, 11(1): 164–169.
48. Kim K, Coh S, Tan LZ, Regan W, Yuk JM, Chatterjee E, Crommie MF, Cohen ML, Louie SG, Zettl A (2012). Raman spectroscopy study of rotated double-layer graphene: misorientation-angle dependence of electronic structure. Phys Rev Lett, 108(24): 246103.
49. Ferrari AC, Meyer JC, Scardaci V, Casiraghi C, Lazzeri M, Mauri F, Piscanec S, Jiang D, Novoselov KS, Roth S, Geim AK (2006). Raman spectrum of graphene and graphene layers. Phys Rev Lett, 97: 187401.
50. Wang YY; Ni ZH, Shen ZX, Wang HM, Wu YH (2008). Interference enhancement of Raman signal of graphene. Appl Phys Lett, 92: 043121.
51. Gupta A, Chen G, Joshi P, Tadigadapa S, Eklund PC (2006). Raman scattering from high-frequency phonons in supported n-graphene layer films. Nano Lett, 6(12): 2667–2673.

52. Eckmann A, Felten A, Mishchenko A, Britnell L, Krupke R, Novoselov KS, Casiraghi C (2012). Probing the nature of defects in graphene by Raman spectroscopy. Nano Lett, 12: 3925–3930.
53. Tuinstra F, Koenig JL (1970). Raman spectrum of graphite. J Chem Phys, 53: 1126–1130.
54. Casiraghi C, Hartschuh A, Qian H, Piscanec S., Georgi C., Fasoli A., Novoselov KS, Basko DM, Ferrari AC (2009). Raman spectroscopy of graphene edges. Nano Lett, 9(4): 1433–1441.
55. Lucchese MM, Stavale F, Ferreira EM, Vilani C, Moutinho MVDO, Capaz RB, Achete CA, Jorio A (2010). Quantifying ion-induced defects and Raman relaxation length in graphene. Carbon, 48(5): 1592–1597.
56. Cançado LG, Jorio A, Ferreira EHM, Stavale F, Achete CA, Capaz RB, Moutinho MVO, Lombardo A, Kulmala TS, Ferrari AC (2011). Quantifying defects in graphene via Raman spectroscopy at different excitation energies. Nano Lett, 11, 3190–3196.
57. Ferrari AC, Robertson J (2000). Interpretation of Raman spectra of disordered and amorphous carbon. Phys Rev B 61: 14095–14107.
58. JF Rodriguez-Nieva, EB Barros, R Saito, Dresselhaus MS (2014). Disorder-induced double resonant Raman process in graphene. Phys Rev B, 90: 235410.
59. Froehlicher G, Berciaud S (2015). Raman spectroscopy of electrochemically gated graphene transistors: Geometrical capacitance, electron-phonon, electron-electron, and electron-defect scattering. Phys Rev B, 91(20): 205413.
60. Vecera P, Chacón-Torres JC, Pichler T, Reich S, Soni HR, Görling A, Edelthalhammer K, Peterlik H, Hauke F, Hirsch A (2017). Precise determination of graphene functionalization by in situ Raman spectroscopy. Nat Commun, 8(1): 15192.
61. Rao CNR, Voggu R (2010). Charge-transfer with graphene and nanotubes. Mater Today, 13 (9): 34–40.
62. Voggu R, Das B, Rout CS, Rao CNR (2008). Effects of charge transfer interaction of graphene with electron donor and acceptor molecules examined using Raman spectroscopy and cognate techniques. J Phys Condens Matter, 20: 472204.
63. Claye A, Rahman S, Fischer JE, Sirenko A, Sumanasekera GU, Eklund PC. (2001). In situ Raman scattering studies of alkali-doped single wall carbon nanotubes. Chem Phys Lett, 333 (1–2): 16–22.
64. Do Nascimento GM, Silva TB, Corio P, Dresselhaus MS (2010). Charge-transfer behavior of polyaniline single wall carbon nanotubes nanocomposites monitored by resonance Raman spectroscopy. J Raman Spectroscopy, 41(12): 1587–1593.
65. Paulus GL, Wang QH, Ulissi ZW, McNicholas TP, Vijayaraghavan A, Shih CJ, Zhong J, Strano MS (2013). Charge transfer at junctions of a single layer of graphene and a metallic single walled carbon nanotube. Small, 9(11): 1954–1963.
66. Voggu R, Rout CS, Franklin AD, Fisher TS, Rao CNR (2008). Extraordinary sensitivity of the electronic structure and properties of single-walled carbon nanotubes to molecular charge-transfer. J Phys Chem C, 112(34): 13053–13056.
67. Huo TT, Yin H, Zhou D, Sun L, Tian T, Wei H, Hu N, Yang Z, Zhang Y, Su YJ (2020). Self-powered broadband photodetector based on single-walled carbon nanotube/GaAs heterojunctions. ACS Sustainable Chem Eng, 8(41): 15532–15539.
68. Rao R, Pierce N, Dasgupta A (2014). On the charge transfer between single-walled carbon nanotubes and graphene. Appl Phys Lett, 105(7): 073115.
69. Cai BF, Yin H, Huo TT, Ma J, Di ZF, Li M, Hu NT, Yang Z, Zhang YF, Su YJ (2020). Semiconducting single-walled carbon nanotube/graphene van der Waals junctions for highly sensitive all-carbon hybrid humidity sensors. J Mater Chem C, 8(10): 3386–3394.
70. Das A, Pisana S, Chakraborty B, Piscanec S, Saha SK, Waghmare UV, Novoselov KS, Krishnamurthy HR, Geim AK, Ferrari AC, Sood AK (2008). Monitoring dopants by Raman scattering in an electrochemically top-gated graphene transistor. Nat Nanotechnol, 3(4): 210–215.
71. Das A, Chakraborty B, Piscanec S, Pisana S, Sood AK, Ferrari AC (2009). Phonon renormalization in doped bilayer graphene. Phys Rev B, 79(15): 155417.

72. Dong X, Fu D, Fang W, Shi Y, Chen P, Li LJ (2009). Doping single-layer graphene with aromatic molecules. Small, 5(12): 1422–1426.
73. Zuo JM, Vartanyants I, Gao M, Zhang R, Nagahara LA (2003). Atomic resolution imaging of a carbon nanotube from diffraction intensities. Science, 300: 1419-1421.
74. Meyer RR, Friedrichs S, Kirkland AI, Sloan J, Hutchison JL, Green MLH (2003). A composite method for the determination of the chirality of single walled carbon nanotubes. J Microscopy, 212(2): 152–157.
75. Hashimoto A, Suenaga K, Gloter A, Urita K, Iijima S (2004). Direct evidence for atomic defects in graphene layers. Nature, 430(7002): 870–873.
76. Sato Y, Yanagi K, Miyata Y, Suenaga K, Kataura H, Iijima S (2008). Chiral-angle distribution for separated single-walled carbon nanotubes. Nano Lett, 8(10): 3151–3154.
77. Qin LC (1994). Electron diffraction from cylindrical nanotubes. J Mater Res, 9(9): 2450–2456.
78. Liu Z, Qin LC (2005). A direct method to determine the chiral indices of carbon nanotubes. Chem Phys Lett, 408(1–3): 75–79.
79. Qin LC (2007). Determination of the chiral indices (n, m) of carbon nanotubes by electron diffraction. Phys Chem Chem Phys, 9: 31–48.
80. Allen C, Zhang C, Burnell G, Brown A, Robertson J, Hickey B (2011). A review of methods for the accurate determination of the chiral indices of carbon nanotubes from electron diffraction patterns. Carbon, 49: 4961–4971.
81. Gao M, Zuo JM, Twesten RD, Petrov I, Nagahara LA, Zhang R (2003). Structure determination of individual single-wall carbon nanotubes by nanoarea electron diffraction. Appl Phys Lett, 82(16): 2703–2705.
82. Liu Z, Zhang Q, Qin LC (2005). Accurate determination of atomic structure of multiwalled carbon nanotubes by nondestructive nanobeam electron diffraction. Appl Phys Lett, 86(19): 191903.
83. Jiang H, Nasibulin AG, Brown DP, Kauppinen EI (2007). Unambiguous atomic structural determination of single-walled carbon nanotubes by electron diffraction. Carbon, 45(3): 662–667.
84. Meyer JC, Kisielowski C, Erni R, Rossell MD, Crommie MF, Zettl A (2008). Direct imaging of lattice atoms and topological defects in graphene membranes. Nano Lett, 8(11): 3582–3586.
85. Robertson AW, Bachmatiuk A, Wu YA, Schäffel F, Rellinghaus B, Büchner B, Rümmeli MH, Warner JH (2011). Atomic structure of interconnected few-layer graphene domains. ACS Nano, 5(8): 6610–6618.
86. Warner JH, Mukai M, Kirkland AI (2012). Atomic structure of ABC rhombohedral stacked trilayer graphene. ACS Nano, 6(6): 5680–5686.
87. Hernandez Y, Nicolosi V, Lotya M, Blighe FM, Sun Z, De S, McGovern IT, Holland B, Byrne M, Gun'Ko YK, Boland JJ, Niraj P, Duesberg G, Krishnamurthy S, Goodhue R, Hutchison J, Scardaci V, Ferrari AC, Coleman, JN (2008). High-yield production of graphene by liquid-phase exfoliation of graphite. Nature Nanotechnol, 3(9): 563–568.
88. PY Huang, CS Ruiz-Vargas, AM van der Zande, Whitney WS, Levendorf MP, Kevek JW, Garg S, Alden JS, Hustedt CJ, Zhu Y, Park J, McEuen PL (2011). Grains and grain boundaries in single-layer graphene atomic patchwork quilts, Nature, 469 (7330): 389–392.
89. Tu X, Manohar S, Jagota A, Zheng M (2009). DNA sequence motifis for strueture-specific recognition and separation of carbon nanotubes. Nature, 460(7252): 250–253.
90. Itkis ME, Perea DE, Jung R, Niyogi S, Haddon RC (2005). Comparison of analytical techniques for purity evaluation of single-walled carbon nanotubes. J Am Chem Soc, 127 (10): 3439–3448.
91. Huang L, Zhang H, Wu B, Liu Y, Wei D, Chen J, Xue Y, Yu G, Kajiura H, Li Y (2010). A generalized method for evaluating the metallic-to-semiconducting ratio of separated single-walled carbon nanotubes by UV-vis-NIR characterization. J Phys Chem C, 114(28): 12095–12098.

92. Tsyboulski DA, Rocha JDR, Bachilo SM, Cognet L, Weisman RB (2007). Structure-dependent fluorescence efficiencies of individual single-walled carbon nanotubes. Nano Lett, 7: 3080–3085.
93. Luo Z, Pfefferle LD, Haller GL, Papadimitrakopoulos F (2006). (n, m) Abundance evaluation of single-walled carbon nanotubes by fluorescence and absorption spectroscopy. J Am Chem Soc, 128(48): 15511–15516.
94. O'Connell MJ, Eibergen EE, Doorn SK (2005). Chiral seleetivity in the charge-transfer bleaching of single-walled carbon-nanotube spectra. Nat Mater, 4(5): 412–418.
95. Li LJ, Nicholas RJ, Chen CY, Darton RC, Baker SC (2005). Comparative study of photoluminescence of single-walled carbon nanotubes wrapped with sodium dodecyl sulfate, surfactin and polyvinylpyrrolidone. Nanotechnology, 16(5): S202.
96. Iakoubovskii K, Minami N, Kim Y, Miyashita K, Kazaoui S, Nalini B (2006). Midgap luminescence centers in single-wall carbon nanotubes created by ultraviolet illumination. Appl Phys Lett, 89(17): 173108.
97. Wang D, Chen L (2007). Temperature and pH-responsive single-walled carbon nanotube dispersions. Nano Lett, 7(6): 1480–1484.
98. Niyogi S, Densmore CG, Doorn SK (2009). Electrolyte tuning of surfactant interfacial behavior for enhanced density-based separations of single-walled carbon nanotubes. J Am Chem Soc, 131(3): 1144–1153.
99. Xie L, Liu C, Zhang J, Zhang Y, Jiao L, Jiang L, Dai L, Liu Z (2007) Photoluminescence recovery from single-walled carbon nanotubes on substrates. J Am Chem Soc, 129: 12382–12383.
100. Ge M, Sattler K (1993). Vapor-condensation generation and STM analysis of fullerene tubes. Science, 260: 515–518.
101. Ge M, Sattler K (1994). Scanning tunneling microscopy of single-shell nanotubes of carbon. Appl Phys Lett, 65: 2284–2286.
102. Odom TW, Huang JL, Kim P, Lieber CM (1998). Atomic structure and electronic properties of single-walled carbon nanotubes. Nature, 391: 62–64.
103. Wilder JW, Venema LC, Rinzler AG, Smalley RE, Dekker C (1998). Electronic structure of atomically resolved carbon nanotubes. Nature, 391(6662): 59–62.
104. Ouyang M, Huang JL, Cheung CL, Lieber CM (2001). Atomically resolved single-walled carbon nanotube intramolecular junctions. Science, 291(5501): 97–100.

Chapter 5
Carbon-Based Heterojunction Broadband Photodetectors

5.1 Introduction

As one type of important energy-conversion devices, photodetector can convert light photons into electrical signals with different working modes, which has been widely used in many fields related to light, including electro-optical displays, optical communication, quantum computing, security monitoring, remote sensing, bio-imaging, defense weapons, and so on. Fox example, the silicon-based photodetectors are commonly used to detect the lights in the UV-visible region, and the near infrared (NIR) lights are detected using the InGaAs-based photodetectors, while HgCdTe are chosen as active materials for mid-IR detectors. For different application conditions, the point or focal plane photodetectors are chosen, however, the capability of converting lights into electrical signals is the most critical whether for point detectors or focal plane detectors. Usually, the larger the photocurrent generated by incident photon excitation is, the higher the performance of photodetectors is. For most commercial photodetectors, the responsivity is lower than 1 A/W, thereby it is a big challenge to obtain high-resolution focal plane photodetectors because low photocurrent results in more complex driver circuit due to smaller photosensitive area. In addition, the detection capability of weak lights is another key parameter for a high performance photodetector. Commonly, the lower the dark current is, the stronger the detection capability is. However, the detection capability cannot be enhanced only through lowering the dark current continuously at room temperature. The fundamental reason for limiting the device performance is the photoactive material itself. Therefore, a tremendous amount of interest in exploring new photoelectric materials has been attracted. In particular, the nanomaterials with unique optoelectronic properties and related nanodevices have been widely investigated with the development of nanotechnology.

As a typical quasi-1D nanomaterial, single-walled carbon nanotube (SWCNT) possesses exceptional electronic, optical, thermal properties with excellent chemical stability due to its unique sp^n hybridization, atom configuration and all atomic

Y. Su, *High-Performance Carbon-Based Optoelectronic Nanodevices*,
Springer Series in Materials Science 319,
https://doi.org/10.1007/978-981-16-5497-8_5

surface [1–3]. SWCNTs exhibit ultrahigh intrinsic mobility (10^5 cm^2/Vs), high absorption coefficient (10^4–10^5/cm) and high thermal conductivity of 6000 W/m K [4–6]. Moreover, the electron transport along the axial direction of SWCNTs is ballistic over several hundred nanometers, enabling high carrier transportation capability when SWCNTs as channels. In addition, semiconducting SWCNTs (sc-SWCNTs) with direct bandgap exhibit a diameter-dependent bandgap feature and broadband optical absorption from UV to mid-infrared (MIR) spectrum range [7]. It is worth noting that the multiple exciton generation (MEG) and multiphoton absorption (MPA) of carrier have also been demonstrated in SWCNTs [8]. The abovementioned features enable SWCNTs as ideal light-harvesting material for novel photoelectronic nanodevices, opening up exciting new possibilities for next-generation high performance photodetectors [9–12].

Another single-atomic layer carbon nanomaterial, graphene, has also been demonstrated to show a great promising potential for applications in nanophononics, photonics, electronics and optoelectronics due to its ultrahigh carrier mobility, unique electronic structures and strong interband transition properties [13–15]. Although monolayer graphene has a low optical absorption ($\sim$2.3%), unique electronic structure with linear dispersion make graphene absorb photons with different energy in the wide region from UV to THz range [16, 17]. Especially, ultrafast carrier dynamics and high mobility of graphene make it suitable to fabricate high-speed photodetectors [18–20]. By combining graphene with other semiconductors, the ultrahigh mobility of graphene and high light absorption of semiconductors can be used simultaneously, resulting in a high optoelectronic conversion performance due to the synergistic effect [21–23].

In this chapter, we simply introduce different working modes and basic parameters of photodetectors. And then, the state-of-art research progresses in SWCNT- and graphene-based photodetectors are systemically summarized. Finally, the current challenges and future perspectives of carbon-based photodetectors are presented.

5.2 Physical Mechanism and Key Parameters of Photodetectors

5.2.1 Physical Mechanism

For a photodetector, the fundamental physical mechanism involved is to convert light signals into the electrical signal (photocurrent or photovoltage) via the photoelectric effect, in which the photogenerated charge carriers are separated, transported through different interfaces, and extracted to external circuits. According to different photoelectric processes, the detecting mechanisms can be classified into five types, including photoconductive effect, photovoltaic effect, photogating effect, bolometric effect and photothermoelectric effect, which corresponds to the photoconductors, photodiodes, phototransistors and thermal detectors, respectively. The abovementioned physical mechanism and related photodetectors will be introduced simply as following.

The photoconductive effect in a photoconductor is the simplest photoelectric process related on the nature of semiconductors, in which the conductivity of semiconductors changes under light illumination due to the formation of photogenerated electron/hole pairs. With a certain bias voltage, the photoexcited electron/hole pairs are separated and then free electrons/holes move oppositely towards the external circuit, resulting in an extra current. The typical I–V curves before and after light illuminations have the same shape except for different slopes, and no photocurrent is generated at zero bias. The photodetectors based on photoconductive effect exhibit large photocurrent, high internal gain, but the response speed and on/off ratio are relatively low.

The photovoltaic effect can be usually observed at p/n junction or Schottky barrier photodiodes, where the asymmetric feature of I–V curve is very clear owing to the rectifying effect. Under light illumination, the built-in electric field at a junction promotes the effective separation of photogenerated electron/hole pairs, and free electrons and holes are transported oppositely towards the external circuit, forming a photocurrent at zero bias. Although photocurrent is usually small in photodiodes, the high detectivity, fast response speed and large bandwidth can be easily obtained due to the lowest dark current at zero bias and the high separate efficiency of photogenerated charge carreries.

The photogating effect is another important photoelectric effect, which occurs in the phototransistor. The device structure is composed of source, drain and gate electrodes, which is relatively more complex than others. The photocurrent is controlled by gate voltage, and the photogenerated electrons or holes are transferred to the channel and another kind of carrier is captured by the external trap centers. Importantly, before the recombination of photogenerated carriers, the electrons or holes can transport several times through the channel, which will give a longer carrier lifetime. Consequently, the phototransistor usually exhibits higher responsivity and larger gain than those of photoconductors. But the phototransistor usually suffers from slower response speed.

For thermal detectors, two physical mechanisms are usually involved: bolometric effect and photothermoelectric effect. In a bolometer, the resistance of the semiconductor changes with the increase of temperature under light illumination. The I–V curves of the bolometers are similar to that of photoconductor, and the responsivity is mainly determined by the temperature coefficient of resistance (TCR) of the semiconductors. Different from the bolometers, the photothermopile based on photothermoelectric effect can work under zero voltage, where a temperature gradient is formed under light illumination, generating a photocurrent across the device owing to high Seebeck coefficient of semiconductors. For CNT-based photothermopiles, they can response to broad electromagnetic spectrum from UV to THz region at room temperature [7, 24, 25].

5.2.2 *Key Parameters*

In order to characterize and compare the detecting performance of photodetectors with different active materials, physical mechanisms and device structures, some commonly used parameters are introduced, including responsivity, detectivity, gain, response speed, bandwidth, external quantum efficiency, etc. The definition and related influencing factors of these key parameters will be introduced briefly as following.

Responsivity (R_λ) is usually defined as the ratio of the photocurrent over the incident light power density, which is widely used to characterize the capability to produce photocurrent for an incident light with given power density and wavelength. The R_λ value can be expressed by Eq. (5.1):

$$R_\lambda = \frac{I_{ph}}{P_{in}} = \frac{et\alpha}{hv} \cdot \frac{\tau_1}{\tau_t} (\mathrm{A\,W^{-1}}) \tag{5.1}$$

where I_{ph} and P_{in} are the photocurrent and the power density of incident light. Sometimes, photovoltage is also used to define the R_λ value. Certainly, the R_λ value not only depends on the absorption coefficient (α) and thickness (t) of the active materials at incident photon energy (hv), but is also strongly influenced by the ratio of the lifetime (τ_l) of minority carriers to the transit time (τ_t) of majority carriers. Theoretically, ultrahigh absorption coefficient (α) and carrier mobility of semiconducting SWCNTs enable the device to exhibit high responsivity (R_λ), which will play an important role in developing ultrahigh pixel photodetection system.

Detectivity (D^*) is another important keyparameter, which is mainly used to characterize the detection capability of weak light for a photodetector. The D^* value is typically calculated using the following formula (5.2):

$$D^* = \frac{\sqrt{A \cdot \Delta f}}{\mathrm{NEP}} \tag{5.2}$$

where A, Δf and NEP are the active area, electrical bandwidth of the photodetectors and noise equivalent power, respectively. When the light density is over the NEP, the photodetector can differentiate an optical signal from noise. If a photodetector has a minimum significance of NEP, the D^* can be simply written as following Eq. (5.3):

$$D^* = \sqrt{\frac{A}{2eI_d}} R_\lambda \tag{5.3}$$

where e and I_d are the elementary charge and dark current, respectively. The detectivity is determined by the responsivity and the noise of photodetector. Obviously, the detection capability of a photodetector can be enhanced by reducing noise or dark current as low as possible and improving the responsivity.

Gain (G) is intrinsic photocurrent amplification, that is, the number of charges collected by electrodes excited by an incident photon. For a photoconductor, G is defined as the number of charges flowing through an external circuit per incident photon, which can be calculated from the wavelength dependent responsivity (Eq. 5.4).

$$G = R_{\lambda} E_{hv} \tag{5.4}$$

where E_{hv} is the energy (eV) of the incident photons. The $\boldsymbol{G}$ can also be calculated by the measured lifetime (τ_{lifetime}) of carrier-recombination and transit time ($\tau_{\text{transit time}}$) (Eq. 5.5)

$$G = \frac{\tau_{lifetime}}{\tau_{transit\ time}} = \frac{\mu V \tau_{lifetime}}{d^2} \tag{5.5}$$

where μ, V and d are the bias voltage, carrier mobility and the channel length, respectively.

Response time (τ) of photodetector is an important parameter for detecting a modulated optical signal, such as optical communication, video imaging, etc. The rise time (τ_r) and the fall time (τ_f) are commonly used to characterize the response speed. The τ_r (or τ_f) value is usually defined as the time interval required for the photocurrent rise (decay) from 10% (90%) to 90% (10%) of its max value. In addition, the **bandwidth** (***Δf***) is widely used to characterize the overall response speed of a photodetector, which is defined as the difference between the upper (f_2) and lower (f_1) light modulation frequencies. The photodetector exhibits a uniform response only in this range ($\Delta f = f_2 - f_1$), but no response outside of this range.

External Quantum Efficiency (*EQE*) describes the number of as-generated photoelectrons per incident photon and is given in percentage. The EQE value is defined as:

$$\text{EQE} = \frac{I_{ph}/e}{P_{in}/hv} \times 100\% = R_{\lambda} \frac{hC}{e\lambda} \times 100\% \tag{5.6}$$

where h, υ and C are the Planck's constant, the incident light frequency and light speed, respectively. Clearly, the *EQE* value is determined by the R_{λ} value of photodetector. Thus, the EQE can be improved by suppressing the carrier recombination and trapping before transporting to the electrodes.

5.3 CNT-Based Photodetectors

As we know, the quantized wave-vector perpendicular to the SWCNT axis results in a unique distribution of density of states (DOS) in k-space per energy level due to its 1D confinement nature. There are several paired sharp features (so-called van

Hove singularities) in the DOS distribution of sc-SWCNTs, representing different the occupied and unoccupied states, respectively. The interband transitions between van Hove singularities can be optically excited or charge recombination. The corresponding S_{11} S_{22} and S_{33} transitions represent the bandgap and sub-bandgaps of SWCNTs, which can be easily distinguished in the absorption spectra. Generally, the sc-SWCNTs have a wide diameter distribution in the range of 0.7–2 nm, the corresponding S_{11} and S_{22} transition energy change in the range of 1.1–0.4 and 2.2–0.7 eV, respectively [9]. Moreover, the electronic transitions in SWCNTs are demonstrated to be sensitive to the polarized light owing to the specific angular momentum in sub-bandgaps. On the other hand, the enhanced Coulomb interaction in the confined 1D channel results in the formation of excitons with larger binding energy, which sometimes dominate the optical absorption of SWCNTs in the visible, NIR and mid-infrared (MIR) spectrum range. Moreover, the absorption coefficient of SWCNTs is as high as 10^4–10^5 cm^{-1} in the NIR-MIR spectrum range, which is at least one order of magnitude larger than that of the traditional IR semiconductor [7]. Importantly, the multiple exciton generation (MEG) and multiphoton absorption (MPA) of carrier occurs effectively in SWCNTs may in principle be used to improve the detecting efficiency of CNT-based photodetectors. The abovementioned unique electronic characteristics and high carrier mobility make sc-SWCNTs be promising materials in the broadband photo detection. In addition, some sc-SWCNTs possess high Seebeck coefficient of $\sim$200 $\mu V\ K^{-1}$, which is several times higher than that of m-SWCNTs (40 $\mu V\ K^{-1}$) [9, 26]. Thus, both mixed SWCNTs and sc-SWCNTs can be used as thermal detectors based on thermoelectric and bolometric effects. In this section, we will introduce different type photodetectors based on pure CNT and CNT heterojunctions. The recent progress will be highlighted and the future perspectives will be also given for this field.

5.3.1 High-Performance SWCNT Photodetectors

Although the photoconductivity of sc-SWCNTs has been experimentally observed at 2001, [27] the photogenerated electrons and holes usually remain in an excitonic state with high diameter-depended binding energy, the excitons need to be separated into free carriers with high mobility of 10^2–10^5 $cm^2\ V^{-1}\ s^{-1}$ using an external large bias or an internal electrical field (p/n junction or Schottky barrier), thereby generating the photocurrent in the external circuit. On the other hand, with the presence of m-SWCNTs, the non-radiative recombination time of charge carriers can be remarkably reduced to 1 ps, dramatically suppressing the separation efficiency of photogenerated carriers in sc-SWCNTs [9, 28]. Thus, much more attention should be paid to the separation of photogenerated carriers in the photodetectors based on pure sc-SWCNTs.

According to the number of active materials, the SWCNT photodetectors can be simply divided into three types: individual sc-SWCNT photodetector, aligned sc-SWCNT array photodetector and sc-SWCNT film photodetector. The earlier numerous studies mainly focus on the investigation of the individual sc-SWCNT nanodevices, [29–31] in which the exciton dynamics and optoelectronic properties of SWCNTs can be well studied. For example, Wang et al. [32, 33] has also demonstrated barrier-free bipolar photodiode based on asymmetric electrode contacts, in which a good ohmic contact can be formed between Y or Sc (low work function) electrode and the conduction band of the sc-SWCNT while Pd (high work function) electrode is used to form an ohmic contact with its valence band. When the channel length is less than the depletion width, the band bend will cover the whole channel of the photodiode, resulting in a large built-in electric field. Thereby photogenerated electron/hole pairs are easily separated and a photocurrent is yielded under illumination [33]. When introducing Au nanoparticles on the SWCNT, the localized surface plasmonic effect demonstrates more than three times photocurrent enhancement in SWCNT photodetector with asymmetric contacts. Moreover, stronger localized plasmon coupling for larger diameter CNTs results in the increase of the enhancement effect [34]. An axe-like plasmonic contact electrode (Pb/Au) has been applied to further enhance the photoelectric response in barrier-free bipolar SWCNT photodiode, and it is found that the polarized incident light parallel and perpendicular to SWCNT can be also effectively utilized using this plasmonic contact electrode, thereby improving the quantum efficiency of SWCNT photodetectors. By compared to the device without a plasmonic structure, the best photocurrent enhancement can be nearly 200 times larger than that without plasmonic structure, simultaneously exhibiting a responsivity of 0.1 mA W^{-1} at 1200 nm light illumination [35]. In addition, by integrating a Fabry-Pérot microcavity on individual SWCNT photodetector, utilizing a concept of the "resonance and off-resonance" cavity, the cavity-integrated chirality-sorted SWCNT self-powered photodetectors at resonance-allowed mode can be used to detect specific target signal. The device exhibits a $\sim$ sixfold enhanced photoelectric conversion and higher suppression ratio of the noise [36].

An alternative approach to effectively separate photogenerated carriers is to fabricate a p/n junction in individual SWCNTs using chemical doping. Abdula et al. [37] developed nearly ideal two-terminal SWCNT photodiodes using spatial doping modulation, in which half of the SWCNT channel was covered by PMMA/ tetracyanoquinodimethane (TCNQ) to enhance p-type character while the polyethylenimine (PEI) covered in another region makes SWCNT be air-stable n-type character. The photodiode displays the doping-level dependent performance and the saturation photocurrent of 1.4 nA under light illuminations. Similar chemical doping has also been using the formation of p-i-n junction in individual SWCNT, [38] triethyloxonium hexachloroantimonate (OA) and PEI doping have been adopted to obtain p- and n- type segments at both ends of a SWCNT, respectively. Under a 1550-nm light illumination, the device exhibits a good photovoltaic effect with an external power conversion efficiency of $\sim 4.2\%$.

Although the photodetectors based on individual sc-SWCNTs have been demonstrated to detect effectively the light signal, the light absorption of single SWCNT is very weak, limiting its practical application for photodetection.

Therefore, sc-SWCNTs array or film was later used to build SWCNT photodetectors. However, the electrical transport of SWCNT has been demonstrated to be strongly influenced by the intertube junctions or defects. To minimize the intertube junctions, aligned sc-SWCNT arrays between electrodes are usually adopted, in which excitons can easily diffuse along each SWCNT to the electrodes. For example, Zhang et al. [39] developed multichannel SWCNT photodiodes with asymmetric contacts by high-frequency AC dielectrophoresis and ultrasonic nanowelding (Fig. 5.1a, b), in which the m-SWCNT needs to be first broken down by applying a large current along the channel. The device generated a photocurrent of 1.54 nA at zero voltage bias under light illumination with a power density of 8.8 W cm^{-2}. Zeng et al. [10] reported a photovoltaic NIR photodetector based on aligned sc-SWCNT arrays using asymmetrically contacting structures (Fig. 5.1c, d). Under 785 nm light illumination, the photodiode exhibits an optimal responsivity of $\sim 6.58 \times 10^{-2}$ A W^{-1} and a detectivity of $\sim 1.09 \times 10^{7}$ Jones (or cm $Hz^{1/2}$ W^{-1}) based on the actual area of SWCNTs.

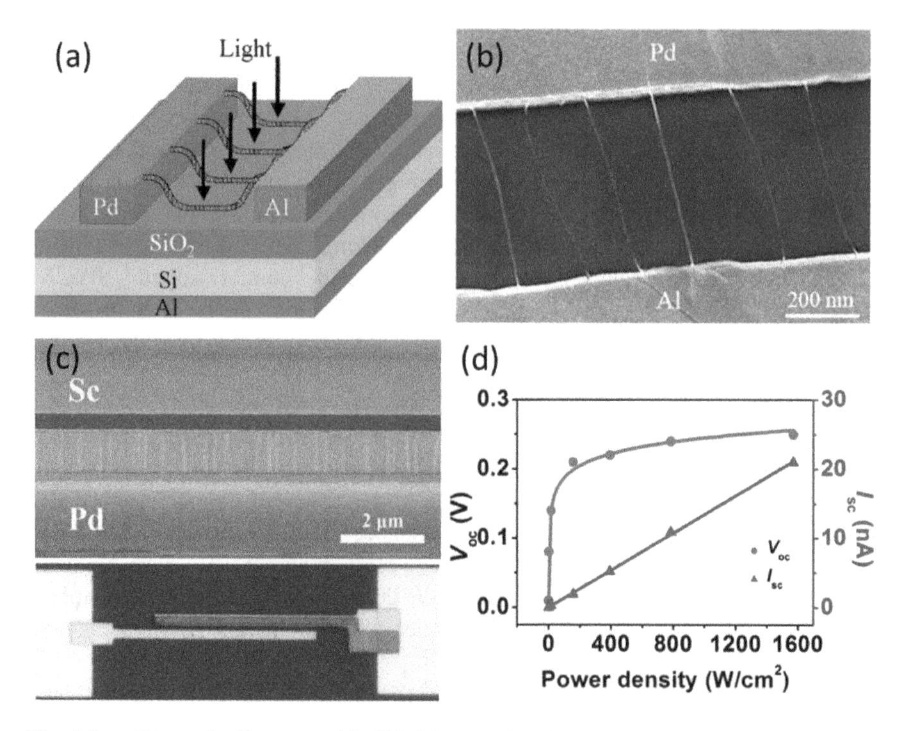

Fig. 5.1 **a** Schematic diagram and **b** SEM image of multichannel SWCNT photodiodes with asymmetric contacts. Reproduced with permission [39]. Copyright 2008, Wiley. **c** SEM image and optical image of SWCNT array photodetector using Sc and Pd as source and drain, respectively. **d** Experimental data and fitted curves of short circuit current and open circuit voltage under 785 nm light illumination with different power density. Reproduced with permission [10]. Copyright 2012, Optical Society of America

Except for SWCNT arrays, the sc-SWCNT network or film has also been used as active materials. Liu et al. [12] demonstrated a high-performance self-powered sc-SWCNT film NIR photodetectors with asymmetric contacts, as shown in Fig. 5.2. Different from common-used photocurrent as the signal, the devices operated at room temperature in voltage mode exhibits broadband response (785–2100 nm), high responsivity and detectivity. Under NIR light (1200–2100 nm) illumination, the largest voltage responsivity and detectivity are as high as 1.5×10^8 V W^{-1} and 1.25×10^{11} Jones occurred at 1800 nm, respectively. Importantly, a wafer scale photodetector array (150 × 150) has been successfully fabricated with high uniformity, which presents a big potential for large-scale fabrication and imager applications. Using similar device structures, Huang et al. [40] demonstrated a random sc-SWCNT network photodetector using asymmetric contacts with plasmonic electrode structure, in which the light absorption and the collect efficiency of photogenerated carriers can be enhanced at the same time. Under 1650 nm light illumination, the detector with plasmonic structure exhibits a responsivity of 7.9 mA $W^{-1}$and a detectivity of 1.46×10^9 Jones for photocurrent mode, while the responsivity and detectivity are 3.5×10^7 V W^{-1} and

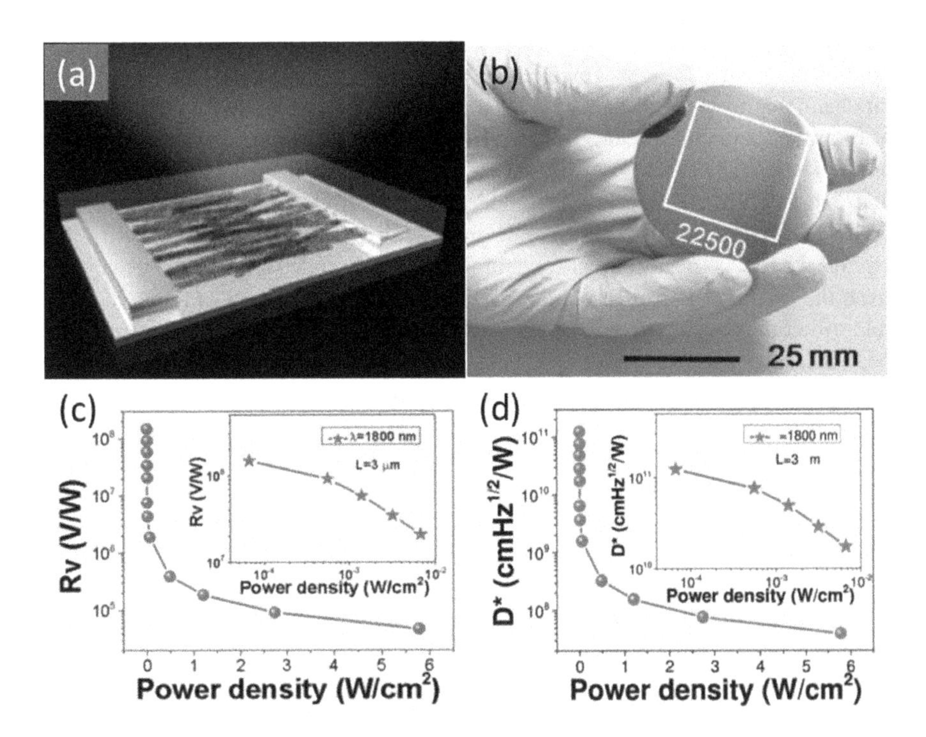

Fig. 5.2 **a** Schematic of self-powered sc-SWCNT film photodetectors with asymmetric contacts. **b** Photograph of a 150 × 150 SWCNT photodetector array on a silicon wafer. **c** Responsivity and **d** detectivity as a function of power density with λ = 1800 nm. Reproduced with permission [12]. Copyright 2016, Wiley

5.79×10^8 Jones for photovoltage mode, respectively. Meanwhile, it is worth noting that the detectivity photocurrent remains constant, while the detectivity at photovoltage mode decreases rapidly with increasing the power density. In addition, since the electronic transitions in sc-SWCNTs are sensitive to the polarized light, the polarimetric photodetector can be constructed using SWCNTs as active materials. Experimental results demonstrated an intrinsic polarization-dependent photocurrent of horizontal aligned SWCNTs film under Vis-NIR light excitation, exhibiting a responsivity of 292.5 mA W^{-1} at 405 nm [41].

It is well known that SWCNTs as ideal IR absorbers can be used as thermal detectors based on bolometric and thermoelectric effects, which will play a very important role in the detection of MIR and THz light. Haddon et al. [7] firstly demonstrated a high IR photoresponse bolometer based on suspended SWCNT film, in which ultrafast relaxation of the photoexcited carriers makes the energy of the incident IR radiation efficiently transfer to the crystal lattice through strong electron–phonon interactions, resulting in the increase of temperature of the SWCNT film. Under 940 nm light illumination, the device in vacuum exhibits the highest responsivity of up to 1000 $VW^{-1}$and a short response time of ~50 ms. The highest temperature coefficient of resistance (TCR) could increase from ~0.7 to 2.5% with reducing the temperature from 330 to 100 K when the thinnest suspended SWCNT film was used. Lu et al. [42] reported a pristine SWCNT film IR bolometer, in which ~80 nm thick SWCNT film is suspended on parallel microchannel arrays on Si substrates. The maximum responsivity is about 250 V W^{-1} at room temperature and the responsivity increases monotonically with decreasing temperature. The response time and detectivity of the detector is ~50 ms and $\sim 4.5 \times 10^5$ Jones at room temperature. Fernandes et al. [43] developed a broadband SWCNT film microbolometer 8×8 array on a thin suspended Si_3N_4 film on Si substrate. The electrical isolation of the suspended microbolometer pixels from the Si substrate can be obtained, and thermal isolation is also improved. This micro-bolometer exhibits a spectral response covering the entire IR range with a peak detectivity of $\sim 1 \times 10^7$ Jones at 100 Hz and short response time of 10 ms at room temperature. Liu et al. [44] reported an ultra-broadband bolometer based on a suspended spider web-like SWCNT film, in which CNTs with wide diameter distribution display a strong absorption spectrum from UV to THz region (375 nm to 118.8 μm). The device achieves a responsivity of ~0.58 A W^{-1} at 375 nm and a fast response time of ~150 μs for all light in vacuum. Even under THz radiation, a responsivity of 11.7 mA W^{-1} and detectivity of 2.31×10^7 Jones can also be expected, giving a promising potential for high-performance terahertz detection at room temperature. It is worth noting that, although many efforts have been made in the development of the SWCNT bolometers, its detectivity is still one to two orders lower than that ($\sim 10^8$ Jones) of commercial bolometers [3].

The photothermoelectric (PTE) detector, namely, photothermopile, is another type of thermal detector, which is a self-powered detector based on the Seebeck effect without external power consumption. Compared to the photonic detectors, these detectors usually show ultra-broadband response from the UV to the THz region, although they suffer from low sensitivity and response speed. Especially, these devices have an obvious advantageous in the THz detection at room temperature. It is well known that SWCNTs exhibit a strong optical absorption and large absorption band with relatively large Seebeck coefficient, making the detectors work in the longwave-IR and the THz range. The SWCNT PTE detector has been developed in the past ten years. St-Antoine et al. [609–613] have demonstrated a position-dependent photovoltage due to the variation of the local Seebeck coefficients in suspended p/n doped SWCNT film thermopile. The thermopile displays an optimum spectral detectivity of 2×10^6 Jones and a time response of 36 ms in the visible and NIR light spectrum, which is comparable to that of SWCNT bolometers. Nanot et al. [45] reported a photodetector based on horizontal aligned SWCNT films onto SiO_2/Si substrate. Under global illumination, the device displays a broadband response from 660 nm to 3.15 μm with a responsivity of $\sim$0.028 V W^{-1} and a response time of <32 μs. Both detailed experimental and theoretical analysis confirms the PTE effect induced photocurrent. The group further developed a visible-MIR broadband photodetector based on horizontally aligned p/n doped SWCNT films, as shown in Fig. 5.3a, b. Under broadband illumination, the device shows a polarization dependent photocurrent owing to the horizontally feature of SWCNT films with the polarization sensitivity, resulting in a max responsivity of $\sim$1 V W^{-1} and a fast response time of $\sim$80 μs [46]. He et al. [25] developed a room-temperature polarization-sensitive THz detector based on highly aligned and ultralong SWCNT array, as shown in Fig. 5.3c, d. The formation of p–n junction is obtained through partial doping of the aligned p-type SWCNT film with benzyl viologen and both type aligned SWCNTs are perpendicular to the current direction. Significant Seebeck coefficient of 75.2 μV K^{-1} for p-type SWCNT film and −71.0 μV $K^{-1}$for n-type SWCNT film make the device exhibit a responsivity of $\sim$2.5 V/W at 3.11 THz under broadband THz irradiation.

5.3.2 CNT/Bulk Semiconductor Heterojunction Photodetectors

In pure SWCNT photodetectors, large exciton binding energy (hundreds of meV) makes it difficult to generate free electron and hole pairs by excitations of first excitons (E_{11}). It usually requires a strong electric field or asymmetric electrical contacts. Therefore, the formation of CNT heterojunctions is considered to be a promising potential solution for high-performance photodetectors. Si and GaAs are typical semiconductors with excellent photoelectric properties, which have been commonly used as commercial photodectectors, such as Si diode and InGaAs diodes. As we know, the key active component in these typical photodetectors is the p/n junction formed by element-doping or metal–organic chemical vapor deposition (MOCVD), which results in high cost, complex and time-consuming process. Moreover, with the development of ultrathin and high-density integrated photodetectors, the light absorption of thin active materials will be a big challenge for

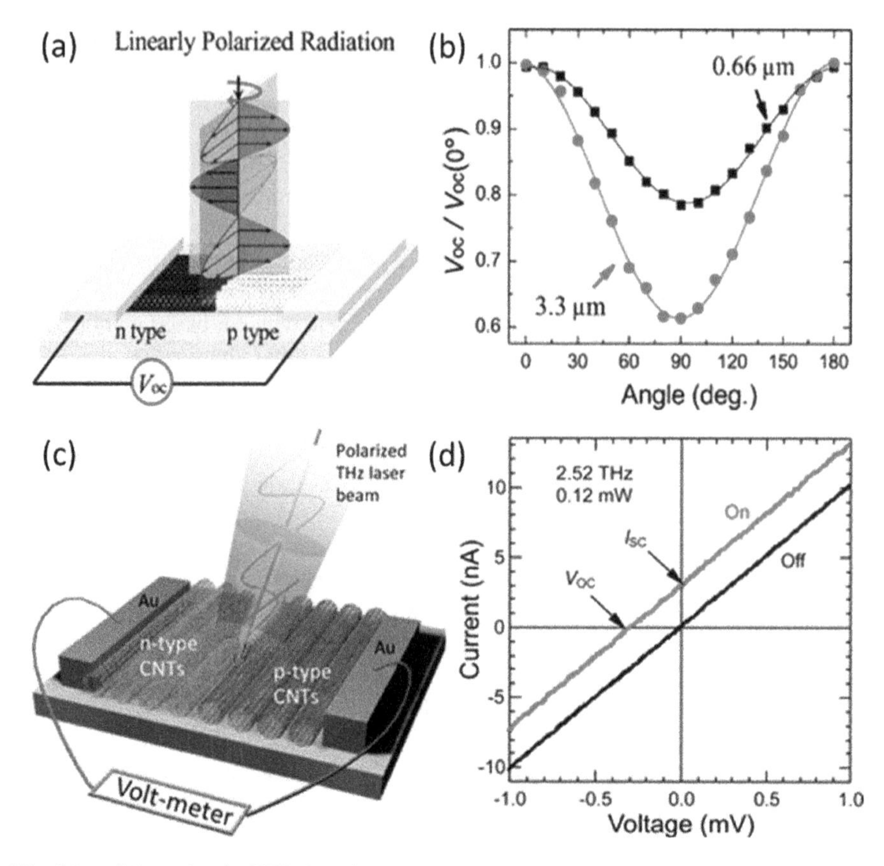

Fig. 5.3 **a** Schematic of a PTE photodetector based on aligned SWCNT film with a p–n junction under illumination of polarized light. **b** Polarization sensitivity at 660 and 3300 nm, respectively. Reproduced with permission [46]. Copyright 2013, American Chemical Society. **c** A schematic diagram of the experimental geometry. **d** Current–voltage characteristics under illumination by a THz beam with a frequency of 2.52 THz, together with that without illumination. Reproduced with permission [25]. Copyright 2014, American Chemical Society

high-performance photodetectors. While the absorption coefficient (10^4–10^5 cm^{-1}) of SWCNTs with high carrier mobility is nearly one order of magnitude higher than that of these bulk semiconductors, and the semiconducting SWCNTs is p-type behavior in air. Therefore, by combing CNTs with the bulk semiconductors, both the unique properties of CNTs and the photoelectric properties of bulk semiconductors can be fully utilized. Meanwhile, the heterostructures can be obtained simply by transferring CNTs on the surface of bulk semiconductors, greatly improving the fabrication efficiency and reducing the cost. More importantly, the atomic-level interface is benefit for the charge separation and transfer due to few disordered structures at the interface between CNTs and bulk semiconductor. Thus, the photodetectors based on the CNT/bulk semiconductor heterojunction are expected to exhibit higher photoelectric performance.

For example, Daniel et al. [47] demonstrated an infrared photoresponse on the vertical CNT array/Si heterostructures, and the NIR to mid-IR photocurrent response was firstly observed in both the cooled and uncooled modes [5800–5804]. After the interface optimization and the use of high quality CNTs, the charge transfer and separation efficiency were enhanced significantly, resulting in good photoelectric performance [48, 49]. An et al. [50] developed a SWCNT film/p-Si Schottky photodetector, in which SWCNT film as the transparent electrode are used to collect the photocurrents which are generated inside the p-Si substrate. The devices exhibit a responsivity of 0.10 A W^{-1} and the noise equivalent power is as low as 1.4×10^{-10} W. Scagliotti et al. [51] reported the SWCNT/Si heterojunction photodetectors with multifinger electrode geometries by dry transferring SWCNTs on n-doped Si substrate. After optimizing the multifinger geometries, the device shows a responsivity of 0.8 AW^{-1}, detectivity of 7×10^{13} Jones, and response time of 7 μs, respectively. The photo response of the SWCNT/Si heterojunction was further investigated under nanosecond pulse laser illumination [52]. Experimental results show that the devices give a fast rise time of 20 ns in photovoltaic mode while the rise time further decreases to less than 10 ns in a photoconductive mode. This fast response is attributed to the formation of Schottky junctions due to the existence of metallic SWCNTs. Salvato et al. [53] also fabricated a high sensitive SWCNT/n-Si heterojunction photodiode using the mixed SWCNTs containing approximately 10% metallic SWCNT and 90% sc-SWCNTs. Under light illumination, the device exhibits a responsivity of $\sim$1 A W^{-1}, detectivity of $\sim 10^{14}$ Jones at zero bias voltage. In addition, when sc-SWCNTs are used to form heterojunctions with Si, only SWCNTs with small diameter as photoactive materials can contribute to photocurrent because the photogenerated electrons in these sc-SWCNTs can be injected into the silicon. The charge transfer process between sc-SWCNTs and Si is determined by the relative positions of the molecular orbitals of the sc-SWCNTs and the band edges of Si [54]. Except for photodetectors, the SWCNT/Si heterojunctions can also be used to fabricate heterogeneous logic circuits. For instance, Kim et al. [55] demonstrated firstly a new kind of photodiode-based logic device using scalable SWCNT/Si heterojunctions (Fig. 5.4), in which the photocurrents with logic outputs can be controlled completely using the logic state of both optical and electronic input. This device exhibits a voltage-switchable photocurrent with the responsivity of more than 1 A W^{-1} and photovoltage responsivity of more than 1×10^5 V W^{-1}. Importantly, the electrical on–off ratios is as high as 2.5×10^5 due to low dark current and short circuit currents, which is difficult to be obtained for mixed-chirality SWCNTs using purely gate voltage. These results pave a new approach to integrating photonics into electronic circuits.

Except for Si, GaAs is another important bulk semiconductor which has been widely used as high efficient space solar cells due to its suitable band gap, high carrier mobility, and superior radiation resistance [56]. Similar to CNT/Si heterojunctions, those composed of CNT and GaAs have also attracted increasing attention since the photoelectric performance of GaAs is better than that of Si. Liang et al. [57] firstly demonstrated systematically the photoelectric properties of

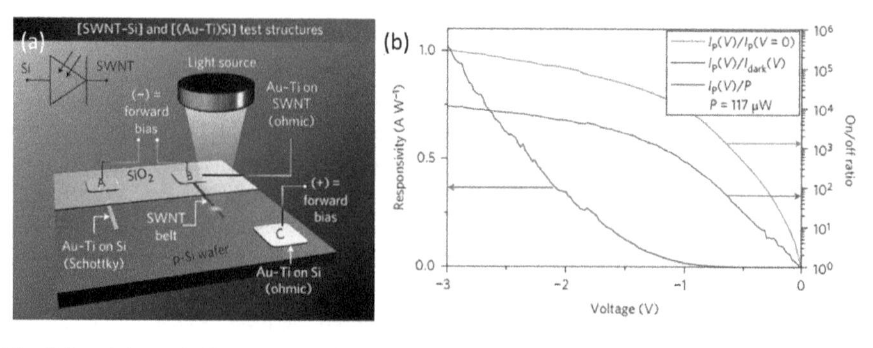

Fig. 5.4 **a** Schematic illustration and **b** the responsivity and on–off ratios of the photodiode-based logic device as a function of the reverse bias. Reproduced with permission [55]. Copyright 2014, Nature Publishing Group

different GaAs/SWCNT heterojunctions. As shown in Fig. 5.5a, b, a strong rectifying behavior is found in the n-GaAs/SWCNT heterojunction, while the contact between p-GaAs and SWCNT results in ohmic-like contact. Under illumination, an obvious photovoltaic effect can be observed in the n-GaAs/SWCNT heterojunctions, laying the groundwork for high-performance self-powered GaAs/SWCNT photodetectors. Li et al. [58] has investigated the photovoltaic response of sc-SWCNT/n-type GaAs heterojunctions, At a low forward bias, the carrier transport dominated by the electron thermionic emission over the energy barrier between the conduction bands of the SWCNT and GaAs, while the hole tunneling through the interband states mainly contributes the current at a low reverse bias. Moreover, the tunneling current plays a significant role under various bias conditions due to the one dimensionality of SWCNTs. Su et al. [59] experimentally demonstrated for the first time a self-power broadband photodetector and solar cell by combining (6, 5) SWCNT-enriched film with n-type GaAs. The device exhibits broadband photoelectric response toward 405–1064 nm due to the light absorption of (6, 5) SWCNT films, breaking though the absorption limits 860 nm of GaAs. As shown in Fig. 5.5c, d, under 405 nm illuminations, high responsivity of 0.274 A W^{-1} and detectivity of 7.6×10^{12} Jones can be achieved with fast response/recover time of 1.41/0.27 ms. One can believe that the responsivity and detectivity could be further improved dramatically by optimizing the SWCNT thickness, minimizing effective active area and interface passivation.

In addition, other CNT/semiconductor heterojunctions have been demonstrated to fabricate highly sensitive photodetectors. For example, Li et al. [60] developed a visible-blind UV photodetector based on vertical sc-SWCNT/ZnO p–n heterojunctions, in which semitransparent large-diameter sc-SWCNT film is covered on surface of the n-ZnO layer. Under UV (370 nm) excitation, the photogenerated electron-hole pairs in the n-ZnO close proximity to the interface can be separated efficiently due to the existence of strong built-in electric field, resulting in a high responsivity of 400 A W^{-1}, ultrahigh detectivity of 3.2×10^{15} Jones and high EQE of 1.0×10^{5}%. Yang et al. [61] demonstrated an heterostructure

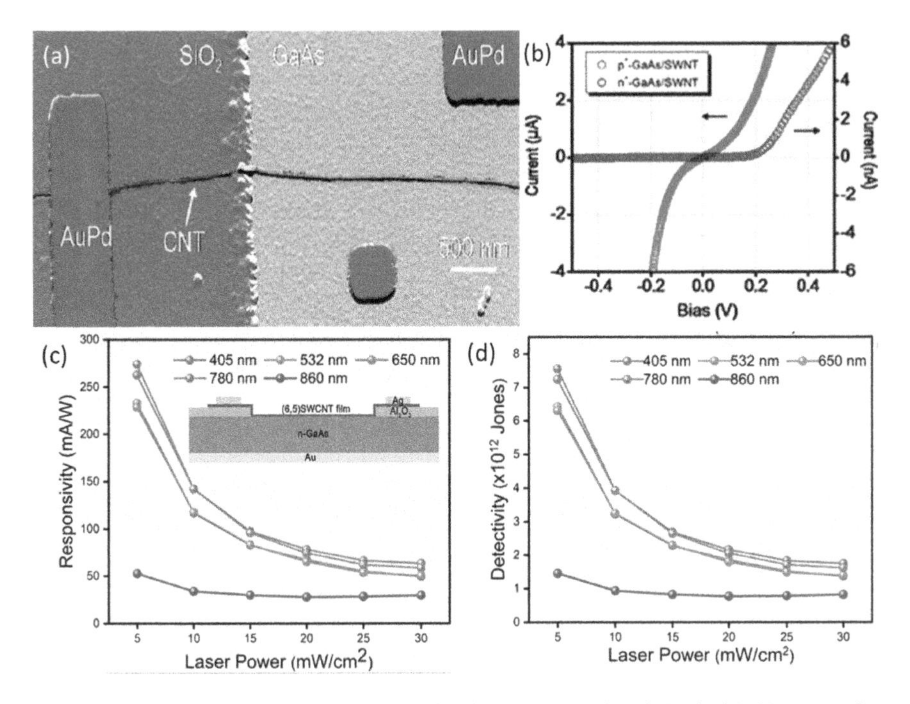

Fig. 5.5 **a** Colored AFM image of the GaAs/SWCNT heterjunction. **b** Typical I−V curves of n- and p-GaAs/SWCNT devices. Reproduced with permission [57]. Copyright 2008, American Chemical Society. **c** Responsivity and **d** detectivity of self-powered (6, 5) SWCNT/GaAs photodetectors as a function of different laser wavelengths. Reproduced with permission [59]. Copyright 2020, American Chemical Society

photodetector using a double-walled CNT film covered on TiO_2 nanotube array with non-ohmic contacts. the photodetector exhibits a high on–off ratio of 10^4 and excellent photoresponse with a responsivity as high as 2467 A W^{-1} under 340 nm irradiation owing to the dimensionality difference effect,. Interestingly, the device exhibits obvious sensitive to visible light (532 nm) where the photon energy is much lower than the band gap of TiO_2.

5.3.3 CNT/Nanomaterial Heterojunction Photodetectors

Compared with bulk semiconductors, low-dimensional semiconducting nanomaterials possess completely different optical and electric properties due to unique quantum size effect and specific surface area. For instance, the physical and chemical properties are more easily influenced by the surface atoms or condiction. Furthermore, the 0D nanomaterials (nanoparticles, nanocrystals or quantum dots (QDs)) exhibit size-depended band gap, and the exciton confinement effect is expected when the particle size is comparable to the exciton Bohr radius, which facilitates novel potential optical applications, such as fluorescence probe, solar cells, photodetectors, and so on [62–64]. Especially, unique multiple exciton generation (MEG) in some semiconducting QDs enables high efficient utilization of high energy photons, producing more photogenerated carriers and thereby corresponding bigger photocurrent. By combining CNTs with semiconducting nanomaterials, high performance photodetectors has been demonstrated. For example, Biswas et al. [65] reported a SWCNT/QD heterojunction phototransistor by decorating CdSe@ZnS core/shell QDs on the surface of p-type SWCNTs, in which a clear type conversion from p-type to n-type can be found due to the charge transfer from QDs to SWCNTs. The device exhibits a sharp electrical photoresponse and enhanced optical Stark effect. Gao et al. [66] anchored PbS QDs on the surface of MWCNTs to fabricated NIR photodetectors, exhibiting a responsivity of 583 mA W^{-1} and detectivity of 3.25×10^{12} Jones. Ka et al. [67] developed a straightforward approach to synthesize SWCNT/PbS QD heterojunctions using pulsed laser deposition (PLD) process, in which PbS QDs can directly atomic contact with surface of SWCNTs. The photoconductive photodetector shows a strong photoresponse as high as 670% and 1350% under 633 and 405 nm light illuminations. Moreover, the size dependent photoresponse in the UV-visible region has been found in the heterojuctions, which is attributed to the MEG induced enhancement effect [68]. Tang et al. [69] demonstrates a SWCNT/PbS QD infrared phototransistor, which exhibits a responsivity of 7.2 A W^{-1} and specific detectivity of 7.1×10^{10} Jones under 1550 nm light illumination. By tuning gate voltage, the responsivity can be further increased to 353.4 A W^{-1}. As we know, the direct chemical decoration or PLD growth of PbS QDs inevitably destroyed the surface of SWCNTs, thereby suppressing the efficient transport of photogenerated carriers. Replacing SWCNTs with double-wall CNTs (DWCNTs), the DWCNT/PbS QD heterojunctions have been formed by PLD anchored QDs on the surface outer nanotube, while the inner nanotubes remains unaffected and can serve efficient pathway for photogenerated charge transportation [70]. Under 250 nm light illuminations, the device exhibits a high responsivity of 230 A W^{-1} and fast response time of 30 μs at an applied voltage of 5 V due to the MEG effect of PbS QDs [71].

In addition, perovskite nanomaterials (QDs or nanowires) have also attracted more and more attention due to the excellent photoelectric properties of perovskite materials in the past few years. The $CsPbI_3$ perovskite QDs were introduced to enhance the light absorption of SWCNTs, sensitizing the creation of free carriers in the SWCNTs. the device exhibits a responsivity of 2.8 A W^{-1} [72]. Spina et al. [73] demonstrates that the photosensitivity of $CH_3NH_3PbI_3$ nanowires can be enhanced significantly by forming the heterojunctions with individual SWCNTs, in which photogenerated holes transfer from the nanowires to nanotube and thereby increase the conductance due to the increase of the carrier number. Under 633 nm laser illuminations ini the nW/mm^2 intensity range, the phototransistor exhibits a high responsivity of 7.7×10^5 A W^{-1} and ultrahigh EQE of 1.5×10^8%. Although only a few photodetectors abovementioned based on CNT-based heterojunctions have been reported, most semiconducting nanomaterials with suitable band alignment can be used to explore high performance CNT/nanomaterial heterojunction photodetectors. However, it is worth noting that the interface control and structural defect suppression need to pay more attention during the improving photoelectric performance.

5.3.4 SWCNT/MoS_2 vdW Heterojunction Photodetectors

Except for bulk semiconductors, nanowires and QDs, the novel van der Waals (vdW) heterojunctions composed of SWCNTs and atomically thin 2D layered semiconductors have also attracted increasing attention due to their unique 1D/2D heterostructures. The physical properties of vdW heterostructures are not only determined by SWCNTs and 2D layered semiconductors, but also tailored by the electron coupling interaction at the interface through element doping or changing the density of electronic states in each material. Thus, these vdW heterojunctions have great potentials in the fabrication of next generation nanoelectronics, flexible electronics, optoelectronics, etc. Among atomically thin 2D layered semiconductors, monolayer MoS_2 is a typical n-type semiconductor with direct band gap of 1.8 eV, [74] which has been demonstrated to exhibit good optoelectronic properties as the visible photodetectors. When n-type monolayer MoS_2 and p-type SWCNTs with excellent NIR absorption are used to build ultrathin vdW heterojunctions, the complementary optical absorption of MoS_2 in the visible region will further improve the performance of SWCNT photodetectors. In addition, the ultrafast charge transfer between SWCNT and MoS_2 can facilitate efficient separation of photogenerated electron/hole pairs at the interface, [75] resulting in high performance of the vdW heterojunction photodetectors. More importantly, the type-II band alignment of the SWCNT/MoS_2 heterojunction can be tailored by changing the diameters of SWCNTs, [76] thereby controlling the interfacial interaction and enhancing the separation efficiency of photogenerated carriers.

Jariwala et al. [77] firstly fabricated a SWCNT/MoS_2 vdW heterojunction phototransistor by combining semiconducting SWCNTs with monolayer MoS_2 nanosheets. The ultrathin nature of vdW heterostructures make both SWCNTs and MoS_2 be modulated via a capacitively coupled gate bias, thereby resulting in a wide tunability of charge transport from a nearly insulating state to a highly rectifying condition. Under 650 nm light illumination, the photodiode exhibits fast response time of <15 μs at $V_g = -40$ V. Nguyen et al. [78] demonstrated a direct synthesis of the SWCNT/MoS_2 1D/2D vdW heterostructures by in-situ chemical vapor deposition of MoS_2 on individual SWCNTs. The phototransistor based on the 1D/2D heterojunctions has been used to detect visible-light, as shown in Fig. 5.6. Experimental results show that the devices could exhibit negative or ambipolar photoresponse depending on the formation of MoS_2 conduction channels along SWCNTs. For pure SWCNT as channel, the phototransistor exhibits typical negative photoresponse due to the electron transfer from MoS_2 to SWCNT. By combining high carrier mobility of SWCNTs as conduction channel with efficient light

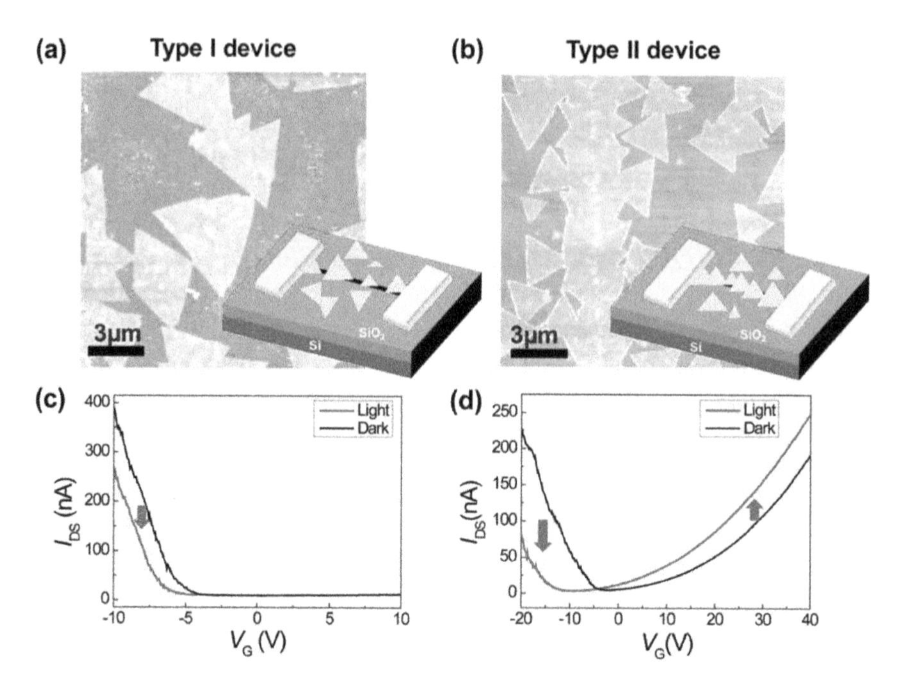

Fig. 5.6 **a–b** Two types of phototransistors based on SWCNT/MoS_2 vdW heterostructures and **c–d** corresponding transfer characteristic curves with and without visible light illuminations. Reproduced with permission [78]. Copyright 2018, WILEY–VCH

absorption of MoS_2 nanosheets, the device still exhibits a high responsivity of $\sim$300 A W^{-1} at zero gate voltage under 532 nm light illumination. Yang et al. [75] demonstrated a high-speed photoinduced memory device based on the MoS_2/SWCNTs network vdW heterostructure (Fig. 5.7), in which an intrinsic ultrafast charge transfer (below 50 fs) occurs at the interface and results in a record program/erase speed of $\sim$32/0.4 ms. Importantly, the device exhibits a record-breaking detectivity of $\sim 10^{16}$ Jones and a high photoresponsivity of 8×10^3 AW^{-1} and under 532 nm laser illumination due to the ultrahigh I_{ph}/I_{dark} ratio.

In the future work, the photoelectric performance of the SWCNT/MoS_2 1D/2D vdW heterojunction phototransistors can be further enhanced from the following aspects. Firstly, single-chiral semiconducting SWCNTs have to be adopted to form uniform type-II band alignment with MoS_2 nanosheets, thereby reducing the recombination probability and suppressing the density of carrier traps which are existed among the mixed chiral SWCNTs. Secondly, free-defect SWCNTs and MoS_2 are used to accelerate the carrier transport, and suitable type-II band alignment is also designed by controlling the band gap of SWCNTs and MoS_2, further enhancing the built-in electric field at the interface. Thirdly, plasmonic nanostructures or interface layer are suggested to applied to promote the interaction between light and vdW heterojuctions or control the charge transfer process between SWCNTs and MoS_2. Finally, the charge transfer dynamic process at the

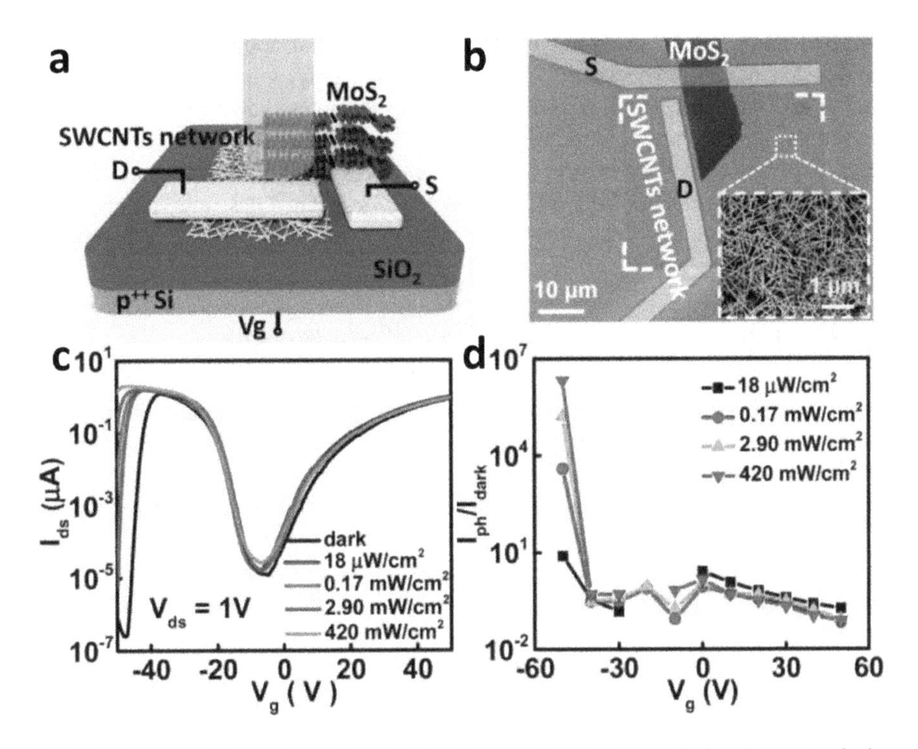

Fig. 5.7 Schematic illustration and optoelectronic properties of the photoinduced memory device based on SWCNT/MoS_2 heterostructures. **a** The schematic and **b** the SEM image of the device. **c** The I_{ds}–V_g curves under laser illumination with different power densities. **d** The ratio of photocurrent (I_{ph}) to dark current (I_{dark}) with different gate voltage. Reproduced with permission [75]. Copyright 2018, WILEY–VCH

interface between SWCNTs and MoS_2 also needs to be investigated in detail, which is very important for the design of higher performance SWCNT/MoS_2 vdW heterojunction photodetectors.

5.4 Graphene-Based Photodetectors

As another important carbon nanomaterial, graphene is composed of single atomic layer of sp^2-bonded carbon atoms arranged in honeycomb lattice, which is the first truly 2D nanomaterial. Importantly, it has unique electronic structure with linear dispersion, vanishing effective mass, high Fermi velocity (ca. 10^6 m s^{-1}), ultrahigh electrical mobility (ca. 2×10^5 cm^2 V^{-1} s^{-1} for free sheet) for electrons and holes, strong interband transition and excellent thermal properties. Unique optical properties enable graphene absorb photons over a broad wavelength range from ultraviolet to THz range. Therefore, these exceptional characteristics of graphene have attracted more and more strong interest in the potential application in nanoelectronics and nano-optoelectronics since 2004. Especially, graphene is expected to enable high-performance photodetection due to its gapless band structure, ultrafast carrier dynamics and high mobility. In this section, we will introduce present state-of-art and recent progress of various graphene-based photodetectors, including pure graphene, graphene/bulk semiconductors, graphene/QDs, and graphene/2D vdW heterostructures.

5.4.1 High-Performance Graphene Photodetectors

Due to the zero band gap feature of graphene, the physical mechanism of photodetectors based on pure graphene is the photogating effect, that is, the graphene device structure is composed of source, drain and gate electrodes. The absorption range can be tuned by changing the Fermi energy of graphene using an external gate field. Xia et al. [19] firstly demonstrated an ultrafast phototransistor based on single- and few-layer graphene, in which the internal fields formed near the metal electrode-graphene interfaces can be used to produce an ultrafast photocurrent, as shown in Fig. 5.8a, b. Moreover, no direct bias voltage between source and drain is needed, which is different from that in the most conventional photodiodes. The device exhibits very wide wavelength detection range, very high bandwidth light detection, and good internal quantum efficiency. However, the weak optical absorption of monolayer graphene and short lifetime of the photo-excited carriers limit the photoresponsivity of photodetectors. In order to enhance the interaction between graphene and light, Furchi et al. [79] designed a graphene-based microcavity photodetector by monolithically integrating graphene with a Fabry-Pérot microcavity, the responsivity can reach to 21 mA W^{-1} because the optical absorption is 26-fold enhanced. Liu et al. [80] demonstrated a dramatic enhancement of the overall quantum efficiency and spectral selectivity of photodetectors by coupling graphene with plasmonic nanostructures. The photocurrent and external quantum efficiency can be enhanced significantly up to 1500% due to enhanced local optical field near the graphene plane. Nevertheless, these approaches restrict photodetection to narrow bands. By integrating a graphene photodetector onto a silicon-on-insulator (SOI) bus waveguide, Gan et al. [81] enhanced the light absorption of graphene and the corresponding detection efficiency without sacrificing the high speed and broad spectral bandwidth. Using a metal-doped graphene junction coupled evanescently to the waveguide, the detector achieves a responsivity of 0.108 A W^{-1} over a wavelength range from 1450 to 1590 nm. Wang et al. [82] demonstrated a graphene/Si heterostructure photodetector by integrating graphene onto a Si optical waveguide on a SOI, in which the evanescent light that propagates parallel to the graphene sheet enable can be absorbed by the waveguide. A responsivity as high as 0.13 A W^{-1} for 2.75 μm light has been firstly obtained at room temperature. Meanwhile, the bias polarity-dependent photocurrent is ascribed to the direct and indirect transitions of carriers in graphene at 1.55 and 2.75 μm, respectively.

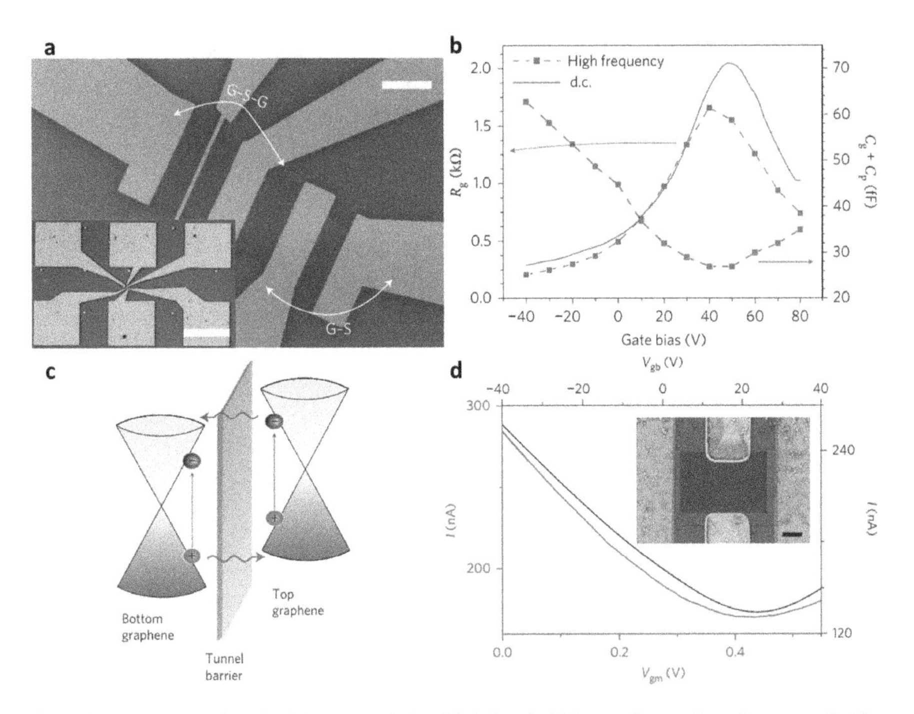

Fig. 5.8 **a** SEM and optical images of the high-bandwidth graphene photodetectors. **b** The measured resistance and total capacitance in the high-frequency domains under different gate biases. Reproduced with permission [19]. Copyright 2009, Nature Publishing Group. **c** Transfer curves of bottom graphene layer (Black) using a silicon backgate and top graphene layer (Red) using the bottom graphene as the gate. **d** Schematic of band diagram and photoexcited hot carrier transport under light illumination. Reproduced with permission [84]. Copyright 2014, Springer Nature

To further improve the maximum responsivity of graphene photodetectors, Kim et al. [83] demonstrated an all-graphene p–n vertical-type tunnelling photodiodes by p type and n-type doping of graphene, which exhibits a responsivity ($0.4 \sim 1.0$ A W^{-1}) and high detectivity ($\sim 10^{12}$ Jones) in broad spectral range from ultraviolet to near-infrared. Such a photoresponse is comparable to or even better than commercially available photodetectors, implying that graphene photodetectors can be used in a variety of possible transparent and flexible optoelectronics applications. Liu et al. [84] designed an ultra-broadband graphene phototransistor based on a double-layer heterostructure (Fig. 5.8c, d), which consists of top monolayer graphene (gate), thin tunnel barrier and bottom monolayer graphene (channel). Under light illumination, photoexcited hot electrons and holes are separated by selective quantum tunneling into opposite graphene layers, minimizing the recombination probability of hot carriers. Moreover, the trapped charges on the top graphene lead to a strong photogating effect on the channel conductance, thereby giving a remarkable responsivity in an ultra-broad spectral range from the visible to the mid-infrared. It is worth noting that ultrahigh responsivity

($>10^3$ A W^{-1}) in the visible region under low excitation power, and higher responsivity can also be achieved by applying a larger bias voltage. In addition, Zhang et al. [85] proposed a broadband graphene photodetector by etching monolayer graphene into a graphene quantum dot-like (GQD)-arrays structures, in which the electron trapping centres are formed due to the introduction of defect midgap states band. The device shows a high responsivity of 8.61 A W^{-1} under 532 nm light illuminations and broadband photoresponse from 532 nm to 10.48 μm.

5.4.2 Graphene/Bulk Semiconductor Heterojunction Photodetectors

Although graphene has been demonstrated to fabricate fast response, broadband photodetectors, it is difficult to further increase significantly the sensitivity due to the weak optical absorption (only ~2.3%) and the gapless feature with short photocarrier lifetime. Naturally, the combination with semiconductors is adopted to enhance the optical absorption, while graphene plays main role in the separation and transport of photogenerated carriers due to its ultrahigh mobility. Especially, bulk semiconductors (Si, GaAs, etc.) offer excellent photoelectric properties, which has been widely used in high efficient photoelectric devices, such as solar cell, photodetectors, and so on. When combing graphene with bulk semiconductors, the unique heterostructures break the traditional restriction in the epitaxial growth, such as lattice mismatch, atomic inter-diffusion, and so on [11]. In addition, the complex growth process can be omitted, thereby reducing the product cost of the photoelectric devices. Therefore, the great potential in these graphene/bulk semiconductor photodetectors have attracted increasing interesting in the past few years.

Chen et al. [86] developed a highly broadband graphene/Si heterostructure photodetector operated in photoconductor mode. Under 632 nm light illuminations, the device exhibits a high responsivity ($>10^4$ A W^{-1}) due to the light absorption of Si, while the responsivity of 0.23 A W^{-1} in the 1550 nm region results from the light absorption of graphene. The significant response is mainly attributed to the prolongation of the ultrashort lifetime of photongenerated carriers by the built-in field in heterostructure. An et al. [21] reported the weak-signal photodetection properties of graphene/Si heterojunctions in the photocurrent and photovoltage modes. Although the low responsivity of 0.435 A W^{-1} is obtained by layer thickening and doping in the photocurrent mode. The photovoltage responsivity exceeding 10^7 V W^{-1} and noise-equivalent-power of ~1 pW $Hz^{-1/2}$ has been demonstrated, indicating excellent weak-light capability of the graphene/Si heterojunctions. To further improve the photoelectric properties, Yu et al. [87] deposited Si QDs on the surface of graphene/Si heterostructures, thereby increasing the built-in potential of the heterojunction and reducing the optical reflection. With the assistance of Si QDs, the graphene/Si Schottky junction photodetector exhibits a high responsivity of ~0.495 A W^{-1} under NIR illuminations and record-short response time of $\leq$25 ns.

Except for bulk Si, GaAs with direct bandgap is also used to form the heterojunctions with graphene in photodetector applications due to its high electron mobility and large optical absorption coefficient (~10^4 cm^{-1}). For example, Wu et al. [88] demonstrated a self-powered photodetector based on graphene/GaAs nanowire heterojunctions, the device presents only a responsivity of 1.54 mA W^{-1} but fast response/recover time of 71/149 μs under 532 nm laser excitation. Luo et al. [89] fabricated a self-powered NIR photodetector by combining monolayer graphene with n-GaAs nanocone array. Under 850 nm illumination, the device exhibits fast response speed of 72/122 μs and a detectivity of 1.83 × 10^{11} Jones. Su et al. [90] further demonstrated the carrier transfer at the interface between graphene and n-type GaAs by monitoring Raman shift of graphene. The as-fabricated self-powered photodetectors exhibits highly sensitive to the visible-NIR light, giving rise to maximum responsivity of 122 mA W^{-1} and detectivity of 4.3 × 10^{12} Jones with quick response/recover time (0.5/0.35 ms), respectively. By inserting h-BN nanosheets between graphene and GaAs. In addition, graphene/Ge heterojunction has also been investigated to detect IR light, [91]. The as-fabricated Schottky junction detector exhibits obvious photovoltaic characteristics, displaying a responsivity of 51.8 mA $W^{-1}$and a detectivity of 1.38 × 10^{10} Jones, respectively. When graphene combines with wide bandgap semiconductors, high-performance UV light photodetector can also be demonstrated [92–94].

5.4.3 Graphene/QD Heterojunction Photodetectors

Colloidal quantum dots (QDs) are known to have display quantum size effect tuning, multiexciton generation (MEG) effect and facile solution processing [95–97]. The quantum size effect allows the effective bandgap and absorption onset to be tuned in a wide range, and the MEG effect enable the creation of two or more electron/hole pairs per absorbed photon, which can be used to enhance the photocurrent in the QD photodectors. However, the QDs exhibit low carrier mobility (10^{-3} −1 cm^2V^{-1} s^{-1}) and the photoconductive gain of corresponding photodetector has been limited to the range of ~10^2 −10^3. By combing graphene with QDs, the mobility-bottleneck of QDs can be overcome effectively due to ultrahigh in-plane carrier mobility of graphene. Thus the 2D/0D heterostructures have been widely investigated using different semiconducting QDs in the past ten years, in which the rGO and CVD-grown graphene are mainly used. As we know, a large number of structural defects and chemical functional groups are formed in the rGO nanosheets during the preparation and reduction process of GO, resulting in the decrease of carrier mobility and the dramatically increase of carrier recombination probability. Although different rGO/QD heterostructures have been successfully used to fabricate novel photodetectors, [98, 99] the photoelectric performance is difficult to be improved significantly. Therefore, the graphene/QD heterojunction photodetectors will be mainly introduced in the following sections.

Konstantatos et al. [100] designed a broadband graphene/QD phototransistor by combining the strong and tunable light absorption of PbS QD thin film with the ultrahigh mobility of mechanically exfoliated graphene (Fig. 5.9). Under light illuminations, the photogenerated electron/hole pairs are separated at the interface and electrons are transferred to the graphene, whereas the holes are trapped in the PbS QDs. The device exhibits firstly an ultrahigh gain of $\sim 10^8$, a responsivity of $\sim 10^7$ A W^{-1} and a specific detectivity of 7×10^{13} Jones. For the practical applications in the future, CVD-grown graphene is more suitable for the fabrication of graphene/QD photodetectors. Zhang et al. [101] demonstrated firstly a reversible phototransistor based on PbS QD decorated CVD-grown graphene, which exhibits responsivity of $\sim 2.8 \times 10^3$ A W^{-1} at the negative gate bias under 400 nm illuminations. Using a similar device architecture, Sun et al. [22] reported the infrared photodetectors based monolayer graphene and PbS QDs, show ultrahigh responsivity of up to 10^7 A/W under 895 light illumination. Such a high photoresponse can be attributed to the field-effect p-type doping in graphene induced by photogenerated electrons in the PbS QDs. Importantly, the capping ligand on the surface of the PbS QDs has been proven to be crucial to the photoelectric performance of the device and needs to be further improved.

Che et al. [102] demonstrated a novel ambipolar NIR vertical photodetector by sandwiching PbS QD film between the graphene and metal electrode, in which the ultrashort channel in the vertical can be provided by controlling the QD layer thickness. Under 808 nm light illuminations, the mixed graphene/QDs as active layer exhibit better photoresponse properties with a responsivity of 1.6×10^4 A/W than pure PbS QDs, which is attributed to the efficient carrier separation and transfer in the mixed graphene/QD layer. In addition, Nian et al. [103] designed a sandwich structural graphene/PbS QDs/graphene by laser shock imprinting, in which graphene with nanoscale wrapping can perfectly cover on the PbS QDs and thereby efficiently collect photogenerated carriers from the PbS QDs. Consequently, this seamless 2D/0D/2D heterostructures can significantly enhance the photoelectric properties of the devices.

Except for PbS QDs, Si QDs as one type of the most important QDs have also attracted much more interests due to their obvious advantageous, such as good photoelectric properties, the abundance and nontoxicity, etc. More importantly, the boron (B) and phosphorus (P) doping enables the localized surface plasmon resonance (LSPR) of Si QDs in the mid-IR (MIR) region, [104, 105] which enables the MIR photodetection of graphene/Si QD heterojunction phototransistors [106]. The photogating effect induced by UV-NIR optical absorption of B-doped Si QDs results in the ultrabroadband UV-MIR photodetection. Moreover, the ultrahigh responsivity of $\sim 10^9$ A W^{-1}), gain of $\sim 10^{12}$) and specific detectivity of $\sim 10^{13}$ Jones have been realized. In addition, III–V semiconductors are also promising candidates for high-performance photodetectors due to their tunable bandgap, excellent photoelectric properties, high mobility and radiation hardness. However, high-quality QDs are hard to be synthesized. Hu et al. [107]

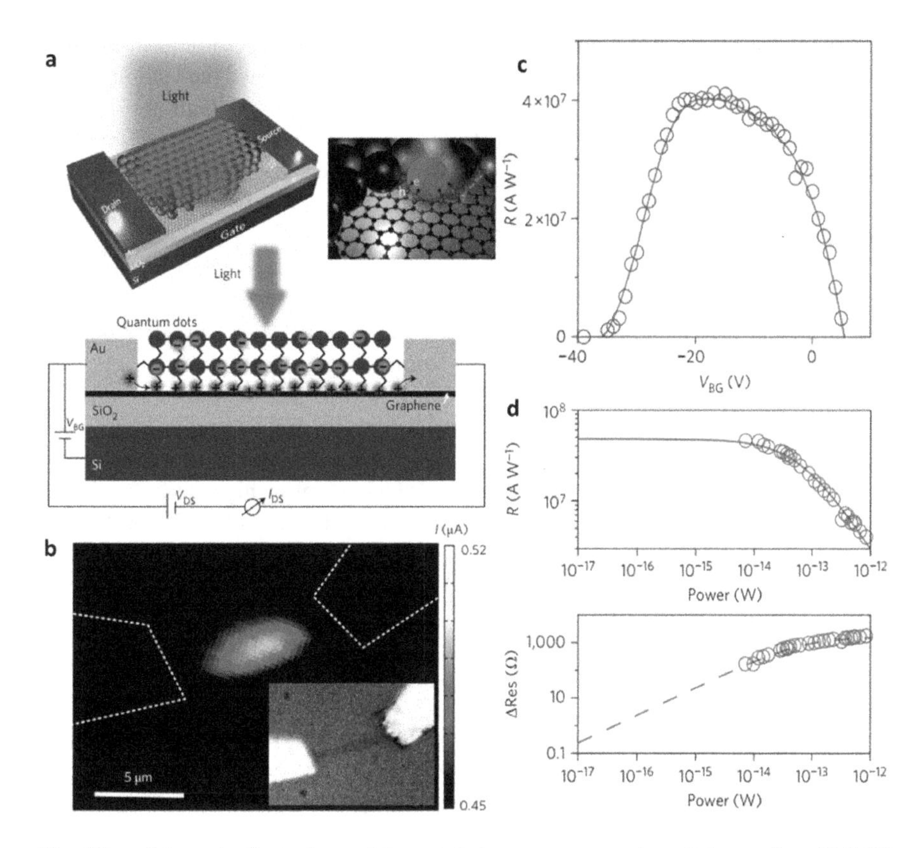

Fig. 5.9 **a** Schematic illustration and **b** spatial photocurrent mapping of the graphene/PbS QD phototransistor under 532 nm laser illumination. **c** Responsivity as a function of back-gate voltage. **d** Responsivity and light-induced resistance change versus laser power. Reproduced with permission [100]. Copyright 2012, Springer Nature

demonstrated a highly sensitive graphene/QD heterojunction photodetector by self-assembling InGaN QDs. which can detect fW light at room temperature without signal/noise processing. Under light illuminations, ultrahigh responsivity of 1.6×10^9 A W^{-1} and specific detectivity of 5.8×10^{14} Jones have been obtained. Such ultrahigh performance is attributed to the photogating gain mechanism, which is consistent with that in abovementioned photodetectors.

Other environmental friendly QDs have also been developed for efficient broadband photodetection by forming graphene/QD heterojunctions. For example, self-doped $Cu_{3-x}P$ QDs exhibit plasmonic resonant absorption in NIR wavelength. Sun et al. [108] reported broadband graphene/QD phototransistors by covering a thin film of $Cu_{3-x}P$ QDs on top of monolayer graphene, which can exhibit an ultrahigh responsivity of 1.59×10^5 A W^{-1} and high photoconductive gain of 6.66×10^5 at 405 nm. Even for 1550 nm illumination, the device still presents a responsivity of 9.34 A W^{-1}. Additionally, the surface ligand on the surface of $Cu_{3-x}P$ QDs plays a key role in determining the charge transfer efficiency from QDs to graphene, which is consistent with that in PbS QDs. Gong et al. [109] demonstrated the viability of the LSPR semiconductor QD/graphene vdW heterostructure for high-performance broadband photodetectors using FeS_2 QDs, in which doped FeS_2 QDs exhibit strong localized surface plasmonic resonance in the wide spectrum ranging from UV to NIR. As a result, the responsivity as high as 1.08×10^6 AW^{-1} can be achieved through optimizing the ligand-exchange process.

In addition, perovskite QDs have also been investigated to fabricate high performance graphene-based photodetectors due to their excellent photoelectric properties and solution-processed advantages. Pan et al. [110] reported a graphene/QD heterojunction phototransistor using formamidinium lead halide perovskite QDs as strong light absorber. Under 520 nm light illuminations, the responsivity and EQE of the device are as high as 1.15×10^5 AW^{-1} and $3.42 \times 10^7\%$, respectively. Inorganic perovskite $CsPbBr_{3-x}I_x$ QDs have also been demonstrated to be form graphene/QD heterojunctions, which display high responsivity of $\sim 10^8$ A W^{-1} and detectivity of $\sim 10^{16}$ jones under 405 nm illumination [111].

5.4.4 Graphene/2D Semiconductor vdW Heterojunction Photodetectors

Hone et al. [112] firstly demonstrated an exceptional high carrier mobility of graphene on 2D hexagonal BN (h-BN) nanosheets due to the reduction of charge scattering sites, indicating that the properties of graphene-based heterostructures are not only determined by the nature of graphene, but also influenced significantly by the electron coupling between graphene and nanosheets. Since then, these van der Waals (vdW) heterostructures consist of graphene and atomically thin-layered semiconductors have been attracting more and more attention in the past few years because of the unique atomic level interfaces and electron-coupling effect, which can be used to build novel ultrathin and flexible photodetectors. Moreover, for artificially stacked vdW heterostructures, the traditional restrictions in the heterostructure growth can be completely broken, such as lattice matching and atomic interdiffusion. Theoretically, various graphene vdW heterostructure photodetectors can be fabricated by simply stacking atomic thin-layered 2D semiconductors with diverse properties on or below the surface of graphene. In the following sections, three type graphene vdW heterostructure photodetectors will be introduced, including graphene/transition metal dichalcogenides (TMDs), graphene/main group metal chalcogenides, and graphene/black phosphorus vdW heterostructures.

1. **Graphene/TMD vdW heterostructures**

Except for graphite, layered TMD materials (such as MoS_2, $MoSe_2$, WS_2, WSe_2, etc.) are also stacked by the weak vdW interactions, enabling the production of monolayer or few layer TMDs using mechanical exfoliation or CVD method.

Especially, most 2D TMDs have layer-dependent bandgaps due to the quantum confinement effects of out-of-plane direction and the symmetry changes, which can provide a large tunability for their optoelectronic properties. By combining graphene with 2D TMD nanosheets, both high carrier mobility of graphene and the excellent optoelectronic properties of semiconducting 2D TMDs will be fully utilized together to fabricate high-performance graphene-based vdW heterojunction photodetectors. In the past few years, these graphene-based photodetectors have been attracting increasing attention.

Zhang et al. [113] demonstrated a graphene/MoS_2 vdW heterostructure phototransistor by transferring graphene onto large-area continuous MoS_2 monolayer. Under light illumination, the photogenerated holes travel from graphene to MoS_2 over the Schottky barrier while photoelectrons are injected into the graphene, resulting in an ultrahigh responsivity of $>10^7$ AW^{-1} and a photogain of ca. 10^8 at $V_g = -10$ V (Fig. 5.10a, b). However, the statistic charge transfer at the interface decreases the barrier height and tunes the Fermi level of graphene, limiting the further improvement of the performance of graphene/MoS_2 vdW heterostructure photodetectors [114]. Li et al. [115] proposed a self-power graphene/h-BN/MoS_2 vdW heterostructure tunneling photodetector by inserting an atomically thin h-BN nanosheet into the graphene/MoS_2 interface (Fig. 5.10c). The interlayer carrier coupling under zero-bias has been substantially blocked while the photogenerated holes can still transport via quantum tunneling, thereby leading to the photocurrent increase over three orders at zero-bias. The photodetector exhibits a high specific detectivity (5.9×10^{14} Jones for white-noise limited detectivity) and a high photoconversion efficiency (EQE > 80%) (Fig. 5.10d). 2D ReS_2 nanosheets with layer-independent direct bandgap can also be form vdW heterostructure photodetector with graphene, presenting an ultrahigh responsivity of 7×10^5 AW^{-1}, a detectivity of 1.9×10^{13} Jones and a fast response time of less than 30 ms [116].

Furthermore, Yu et al. [117] demonstrated highly efficient photocurrent generation in the photodetectors based on graphene/MoS_2/graphene sandwich heterostructures (Fig. 5.11a, b). The asymmetric potential in the top graphene/MoS_2 and bottom graphene/MoS_2 junction results in an apparent photovoltaic effect under 514 nm laser illuminations. Moreover, the photocurrent can be easily modulated by the electric field of an external gate, thereby achieving a maximum EQE of 55% and internal quantum efficiency up to 85%. Similar physical mechanism is also observed in the hBN/graphene/WS_2/graphene/hBN sandwich heterostructures, reaching an extrinsic quantum efficiency of 30% [118]. Massicotte et al. [119] also fabricated a high-speed photodetector based on graphene/trilayer WSe_2/graphene vdW heterostructures encapsulated in hexagonal h-BN (Fig. 5.11c), which presents a photoresponse time as short as 5.5 ps (Fig. 5.11d).

Except for the graphene/TMD/graphene heterostructures, novel P/graphene/N vdW heterostructures have also been used to fabricate by inserting graphene into the 2D p–n heterostructures, reducing the charge traps and improving the charge transport in the vdW heterojunctions. For example, Long et al. [120] fabricated a high-performance broadband (400–2400 nm) photodetector based on the MoS_2/graphene/WSe_2 vdW heterostructures (Fig. 5.12a, b), Under visible light

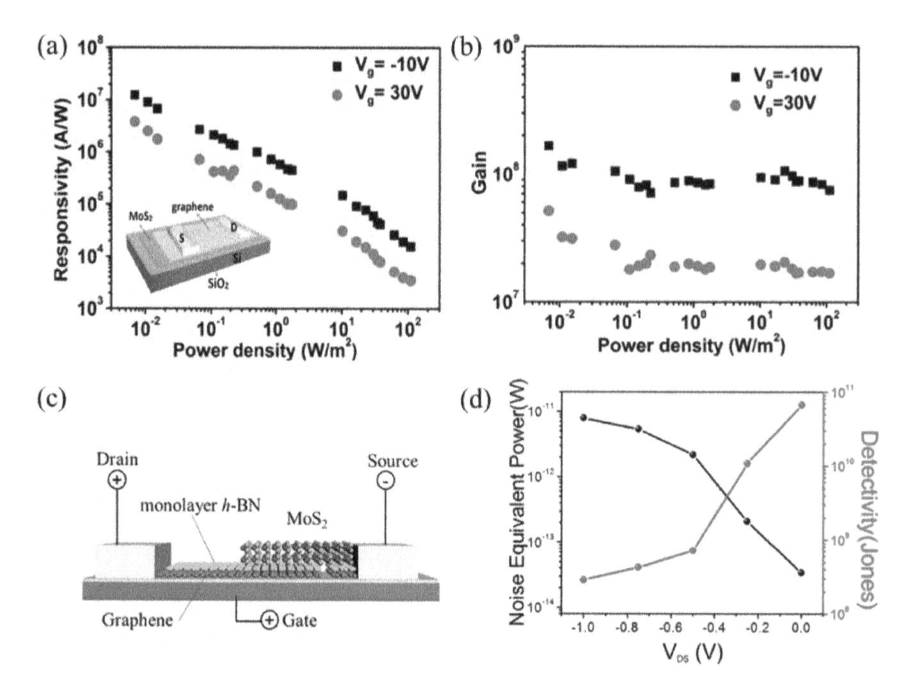

Fig. 5.10 **a** Photoresponsivity and **b** photogain for the graphene/MoS_2 vdW heterostructure photodetector. Reproduced with permission [113]. Copyright 2014, Springer Nature. **c** Schematic of the MoS_2/h-BN/graphene vdW heterostructure device. **d** The measured noise equivalent power (NEP) and specific detectivity at different voltage bias. Reproduced with permission [115]. Copyright 2019, Elsevier

illuminations, the measured responsivity and calculated specific detectivity are as high as 10^4 A W^{-1}, 10^{15} Jones, respectively. Even for NIR lights, the responsivity and specific detectivity can still be maintained at several A W^{-1} and $\sim 10^{11}$ Jones. It is demonstrated that the graphene interlayer plays a critical role in realizing both high detectivity and the responsivity in the UV–vis-NIR photodetectors based on hBN/$MoTe_2$/graphene/SnS_2/hBN vdW heterostructures (Fig. 5.12c, d) [121]. Importantly, the device exhibits a high specific detectivity is up to 1.1×10^{13} Jones under 1064 nm NIR illumination at room temperature, which is larger than that of commercial InGaAs NIR photodetectors operated at extremely low temperatures. Moreover, the specific detectivity value still is up to 1.06×10^{11} Jones at 1550 nm excitation, which can be comparable with commercial short-wave IR photodetectors.

2. **Graphene/main group metal chalcogenides heterostructures**

Similar to 2D TMDs, some main group metal chalcogenides with layered structures have also been formed graphene vdW heterostructures, including group IIIA metal (Ga, In) and group IVA metal (Ge, Sn) chalcogenides. For example, the graphene/GaSe vdW heterostructure photodetector can exhibit a high responsivity

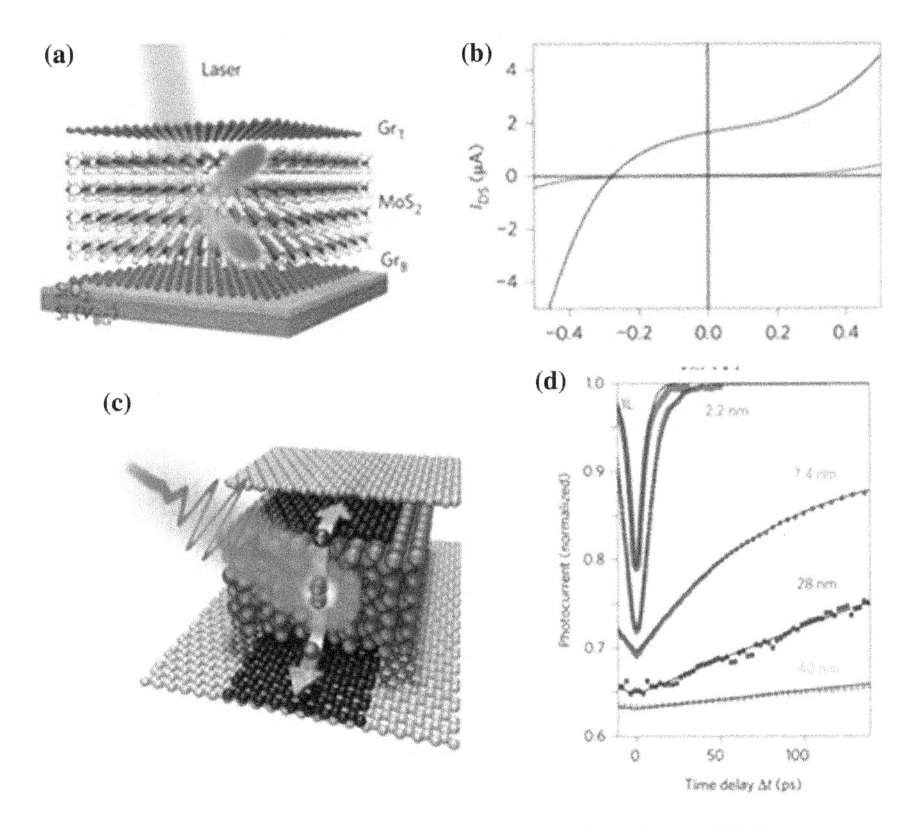

Fig. 5.11 **a** Schematic illustration of the graphene/MoS_2/graphene vdW heterostructure. **b** Current–voltage curves with and without illumination. Reproduced with permission [117]. Copyright 2013, Springer Nature. **c** Schematic illustration of photoexcited carrier dynamics and **d** Time-resolved photocurrent in an hBN/G/WSe_2/G/hBN heterostructure. Reproduced with permission [119]. Copyright 2016, Springer Nature

of $\sim 3.5 \times 10^5$ A/W and a detectivity of 1.1×10^{10} Jones under 532 nm light illumination, and the photoconductive gain is up to 10^7 [122]. Similarly, the graphene/InSe vdW heterostructure photodetectors can also be demonstrated, in which graphene can modify the difference of the work function between InSe and Au electrodes. Under visible light illuminations, the device presents a gate-tuned response time of 100–310 μs and a high responsivity of 60 A W^{-1} [123]. Luo et al. [124] also proposed a InSe/graphene vdW heterostructure photodetector to detect visible-NIR light (400–1000 nm), giving a responsivity of 940 A W^{-1} at 532 nm. Mudd et al. [125] designed a broad-band photodetector based on vertical and planar graphene/n-InSe/graphene heterostructures by mechanical contact. For vertical heterostructures, the top graphene as electrode and broad-band optical window enables an efficient extraction of photogenerated carriers in n-InSe, resulting in a spectral response to visible-NIR spectrum with a high responsivity of ca. 10^5 A W^{-1} and specific detectivity of 10^{13} Jones under 633 nm light illuminations.

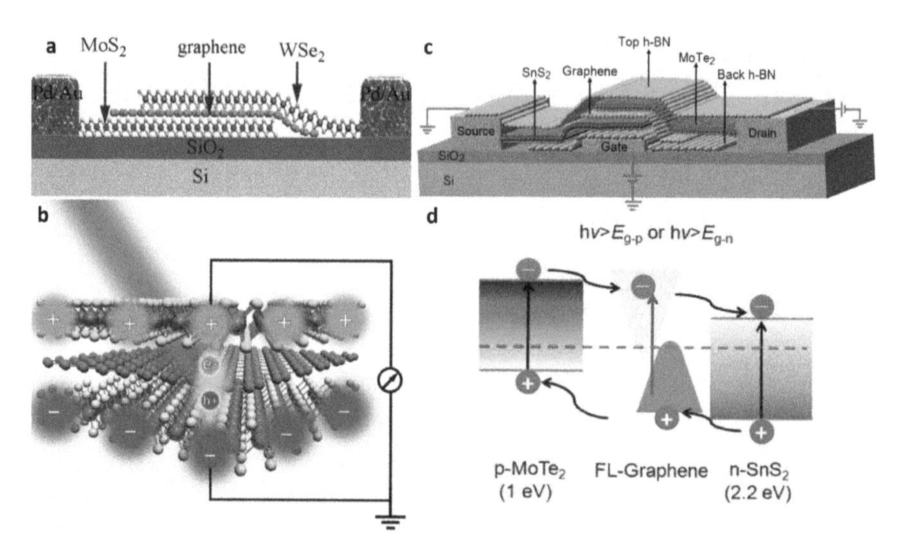

Fig. 5.12 **a** Schematic side view of the MoS_2/graphene/$MoSe_2$ vdW heterostructure and **b** the photovoltaic photodetector based on this vdW structure. Reproduced with permission [120]. Copyright 2016, American Chemical Society. **c** Schematic hBN/$MoTe_2$/graphene/SnS_2/hBN device configuration. **d** Schematic band diagram and photoexcited carriers transport. Reproduced with permission [121]. Copyright 2019, WILEY–VCH

Furthermore, 2D tetradymites (Bi_2Se_3, Bi_2Te_3) have also attracted remarkable attention in the infrared optoelectronic devices due to their unique gapless surface states (so-called topological insulators) and small band gaps. Kim et al. [126] demonstrated a room-temperature IR photodetector based on a graphene/Bi_2Se_3 heterostructure, presenting broadband detection from mid-IR to NIR. Moreover, the device exhibits a record-high responsivity of 8.18 A W^{-1} for 1.3 μm NIR light and the response time is as fast as 4 μs. Qiao et al. [127] reported the formation of graphene/Bi_2Te_3 heterostructure by growing Bi_2Te_3 nanoplates on the top of graphene. The device also has the capability for broadband photodetection from visible to NIR wavelength with a photoresponsivity of 35 A W^{-1} at 532 nm.

3. **Graphene/black phosphorus heterostructures**

Except for graphene, black phosphorus (BP) is another typical 2D layered element semiconductor with layer-dependent direct bandgap from 0.3 to 2.0 eV [128, 129]. In particular, the carrier mobility of 10^4 cm^2 V^{-1} s^{-1} facilitates high-performance broadband photodetectors [130]. Moreover, the strong interaction between a lone pair of electrons at each P atom and out of-plane atoms, which makes BP nanosheets exhibit higher out-of-plane conductivities [131]. Although the degradation of BP nanosheets happens easily in the air condition due the existance of oxygen and water molecules, it can be suppressed dramatically by covering graphene on the top of BP nanosheets [132]. The device performance of vertical graphene/BP vdW heterostructures has also demonstrated to be determined by out-of-plane transport through the vdW heterostructures. Therefore, the photodetectors based on these graphene/BP vdW heterostructures have attracted increasing attention.

For example, Liu et al. [132] demonstrated firstly a high-performance graphene/BP vdW heterostructure phototransistor with stable broadband photoresponse from the visible to IR wavelengths. The transfer of photogenerated carriers from BP to graphene reduces significantly the Schottky barrier between BP and Au electrodes, thereby resulting in an efficient photocurrent extraction with a responsivity of $\sim 3.3 \times 10^3$ A W^{-1} at 1550 nm. The corresponding photoconductive gain is up to 1.13×10^9 and the response time is about 4 ms. Zhou et al. [133] also designed similar graphene/BP vdW heterostructure phototransistors, in which the device exhibits a responsivity of 7.7×10^3 A/W under 360 nm light illumination due to the diversity of BP nanosheets. By transferring multi-layer graphene flake on the top of multi-layer BP flake, a graphene/BP vdW heterostructure photodetector has also been fabricated, and the device presents a high responsivity of 55.75 A W^{-1} at 655 nm, 1.82 A W^{-1} at 785 nm and 0.66 A W^{-1} at 980 nm because of the dramatically decrease of the recombination probability of photoexcited electron/hole pairs, In addition, to further improve the performance of photodetectors, it is important to block significantly dark current fluctuations in the vdW heterostructures. 2D hBN nanosheets as a tunneling barrier has been successfully used to suppress dark current by inserting into graphene/2D semiconductor vdW heterostructures, which can also be used to increase the specific detectivity [134].

5.5 Summary and Outlook

In summary, we summarize systematically the state-of-the-art research progress of broadband photodetectors based on carbon pure SWCNT, pure graphene and related vdW heterostructures. On the whole, with development of carbon nanomaterials, significant progress in the carbon-based photodetectors has been achieved, and the photoelectric performance of carbon-based photodetectors have been improved remarkably in the past ten years. More importantly, these carbon-based devices exhibit much more excellent performance than traditional commercial photodetectors, implying huge potential for future application in security monitoring, photoelectric imaging, optic communication, and so on. However, we have to say that carbon-based photodetectors are still at initial stage, additional challenges have to be addressed before the full commercial potential. Especially, both high-quality carbon nanomaterials and basic device physical mechanism still need to be further revealed deeply, such as single-chirality SWCNTs, clean atomic interface, light-mater interaction, exciton dissociation, interface carrier dynamic process, vdW heterojunctions-induced electron coupling effect, and so on. To further improve the performance of carbon-based broadband photodetectors, much more efforts need to be made in the following aspects in the future work.

Firstly, high-quality carbon nanomaterials with specific properties are one of key parameters during the fabrication of high-performance photodetectors. Especially, high purified sc-SWCNTs should be firstly addressed, while the presence of metallic SWCNTs results in the substantial reduction of the non-radiative charge recombination time, significantly inhibiting the photoresponse sensitivity and speed of photodetectors. Thus it is very critical to remove metallic SWCNTs from sc-SWCNTs or sc-SWCNT films. It will be better if pure single-chirality sc-SWCNTs with suitable bandgaps can be used, because numerous intertube junctions and the tube-tube interactions with different chiralities will trap the excitons or photogenerated carriers, thereby decrease the efficiency of free carrier collection. Therefore, the preferential growth or high efficient separation of single-chirality sc-SWCNTs is still an important issue in the future. On the other hand, large-area monolayer or few-layer graphene with high crystallinity is also highly desired, which enables efficient charge transport in graphene-based photodetectors. In the heterostructures, the cleanliness of the interface is an important factor to influence the charge transfer process and the device performance, requiring that the formation process of the vdW heterostructures needs to be further investigated systematically for both SWCNTs and graphene.

Secondly, some basic device mechanisms needs to be further understood deeply, such as the Fermi level regulation, the light-mater interaction, the exciton dissociation in sc-SWCNTs, carrier dynamic process, etc. Especially, SWCNTs and graphene possess unique electron band structures (van Hove singularity, Dirac cones), which enable more flexible ways to regulate the Fermi level of sc-SWCNTs and graphene. For example, chemical doping, photo-doping, metal ion or noble metal nanoparticle modification can be used to change the Fermi level of both sc-SWCNTs and graphene. Meanwhile, if the Fermi level in sc-SWCNTs is moved into one of the van Hove singularities, more light will be allowed to pass through. For the vdW heterojunctions photodetectors, the interface charge dynamics and the electron coupling effect must be systematically and deeply studied by time-resolved ultrafast spectroscopy. On the other hand, the detailed charge transfer or energy transfer process as well as their competitive relationship at the interface is also fundamental work to understand the working mechanism of photodetectors. For the vdW heterojunctions composed of nanomaterials, the properties of heterojunctions is influenced by both each materials and the electron-coupling interaction, new approaches should be applied to enhance the light absorption of vdW heterostructures, such as external field, plasmonic nanostructures, etc.

References

1. Liu L, Han J, Xu L, Zhou J, Zhao C, Ding S, Shi H, Xiao M, Ding L, Ma Z, Jin C (2020). Aligned, high-density semiconducting carbon nanotube arrays for high-performance electronics. Science, 368(6493): 850–856.
2. Avouris P, Freitag M, Perebeinos V (2008). Carbon-nanotube photonics and optoelectronics. Nature Photonics, 2(6): 341–350.

3. He X, Léonard F, Kono J (2015). Uncooled carbon nanotube photodetectors. Adv Optical Mater, 3(8): 989–1011.
4. Liu Y, Wang S, Peng LM (2016) Toward high-performance carbon nanotube photovoltaic devices. Adv Energy Mater 6(17): 1600522.
5. Peng, LM, Zhang Z, Wang S (2014). Carbon nanotube electronics: recent advances. Mater Today, 17(9): 433–442.
6. Yang L, Wang S, Zeng Q, Zhang Z, Peng LM (2013). Carbon nanotube photoelectronic and photovoltaic devices and their applications in infrared detection. Small, 9: 1225–1236.
7. Itkis ME, Borondics F, Yu A, Haddon RC (2006). Bolometric infrared photoresponse of suspended single-walled carbon nanotube films. Science, 312(5772): 413–416.
8. Wang S, Khafizov M, Tu X, Zheng M, Krauss TD (2010). Multiple exciton generation in single-walled carbon nanotubes. Nano Lett, 10(7): 2381–2386.
9. Richter M, Heumüller T, Matt GJ, Heiss W, Brabec CJ (2017). Carbon photodetectors: the versatility of carbon allotropes. Adv Energy Mater, 7(10): 1601574.
10. Zeng Q, Wang S, Yang L, Wang Z, Pei T, Zhang Z, Peng LM, Zhou W, Liu J, Zhou W, Xie SS (2012). Carbon nanotube arrays based high-performance infrared photodetector. Optical Mater Express, 2(6): 839–848.
11. Yang L, Wang S, Zeng Q, Zhang Z, Peng LM (2013). Carbon nanotube photoelectronic and photovoltaic devices and their applications in infrared detection. Small, 9(8): 1225–1236.
12. Liu Y, Wei N, Zeng Q, Han J, Huang H, Zhong D, Wang F, Ding L, Xia J, Xu H, Ma Z, Qiu S, Li QW, Liang XL, Zhang ZY, Wang S, Peng LM (2016). Room temperature broadband infrared carbon nanotube photodetector with high detectivity and stability. Adv Optical Mater, 4(2): 238–245.
13. Bonaccorso F, Sun Z, Hasan TA, Ferrari AC (2010). Graphene photonics and optoelectronics. Nature Photonics, 4(9): 611–622.
14. Bao Q, Loh KP (2012). Graphene photonics, plasmonics, and broadband optoelectronic devices. ACS Nano, 6(5): 3677–3694.
15. Zhang Y, Liu T, Meng B, Li X, Liang G, Hu X, Wang QJ (2013). Broadband high photoresponse from pure monolayer graphene photodetector. Nature Commun, 4(1): 1–11.
16. Vicarelli L, Vitiello MS, Coquillat D, Lombardo A, Ferrari AC, Knap W, Polini M, Pellegrini V, Tredicucci A (2012). Graphene field-effect transistors as room-temperature terahertz detectors. Nature Mater, 11(10): 865–871.
17. Cai X, Sushkov AB, Suess RJ, Jadidi MM, Jenkins GS, Nyakiti LO, Myers-Ward RL, Li S, Yan J, Gaskill DK, Murphy TE (2014). Sensitive room-temperature terahertz detection via the photothermoelectric effect in graphene. Nature Nanotechnol, 9(10): 814–819.
18. Pospischil A, Humer M, Furchi MM, Bachmann D, Guider R, Fromherz T, Mueller T (2013.) CMOS-compatible graphene photodetector covering all optical communication bands. Nature Photonics, 7(11): 892–896.
19. Xia F, Mueller T, Lin YM, Valdes-Garcia A, Avouris P (2009). Ultrafast graphene photodetector. Nature Nanotechnol, 4(12): 839–843.
20. Mueller T, Xia F, Avouris P (2010). Graphene photodetectors for high-speed optical communications. Nature Photonics, 4(5): 297–301.
21. An X, Liu F, Jung YJ, Kar S (2013). Tunable graphene–silicon heterojunctions for ultrasensitive photodetection. Nano Lett, 13(3): 909–916.
22. Sun Z, Liu Z, Li J, Tai GA, Lau SP, Yan F (2012). Infrared photodetectors based on CVD-grown graphene and PbS quantum dots with ultrahigh responsivity. Adv Mater, 24 (43): 5878–5883.
23. Koppens FH, Mueller T, Avouris P, Ferrari AC, Vitiello MS, Polini M (2014). Photodetectors based on graphene, other two-dimensional materials and hybrid systems. Nature Nanotechnol, 9(10): 780–793.
24. St-Antoine BC, Ménard D, Martel R (2009). Position sensitive photothermoelectric effect in suspended single-walled carbon nanotube films. Nano Lett, 9(10): 3503–3508.

25. He X, Fujimura N, Lloyd JM, Erickson KJ, Talin AA, Zhang Q, Gao W, Jiang Q, Kawano Y, Hauge RH, Léonard F (2014). Carbon nanotube terahertz detector. Nano Lett, 14 (7): 3953–3958.
26. Erikson KJ, He X, Talin AA, Mills B, Hauge RH, Iguchi T, Fujimura N, Kawano Y, Kono J, Léonard F (2015). Figure of merit for carbon nanotube photothermoelectric detectors. ACS Nano, 9(12): 11618–11627.
27. Fujiwara A, Matsuoka Y, Suematsu H, Ogawa N, Miyano K, Kataura H, Maniwa Y, Suzuki S, Achiba Y (2001). Photoconductivity in semiconducting single-walled carbon nanotubes. Japan J Appl Phys, 40(11B): L1229.
28. Lauret JS, Voisin C, Cassabois G, Delalande C, Roussignol P, Jost O, Capes L (2003). Ultrafast carrier dynamics in single-wall carbon nanotubes. Phys Rev Lett, 90(5): 057404.
29. Qiu X, Freitag M, Perebeinos V, Avouris P (2005). Photoconductivity spectra of single-carbon nanotubes: Implications on the nature of their excited states. Nano Lett, 5 (4): 749–752.
30. Freitag M, Martin Y, Misewich JA, Martel R, Avouris P (2003). Photoconductivity of single carbon nanotubes. Nano Lett, 3(8): 1067–1071.
31. Kumamoto Y, Yoshida M, Ishii A, Yokoyama A, Shimada T, Kato YK (2014). Spontaneous exciton dissociation in carbon nanotubes. Phys Rev Lett, 112(11): 117401.
32. Ding L, Wang S, Zhang Z, Zeng Q, Wang Z, Pei T, Yang L, Liang X, Shen J, Chen Q, Cui R (2009). Y-contacted high-performance n-type single-walled carbon nanotube field-effect transistors: scaling and comparison with Sc-contacted devices. Nano Lett, 9(12): 4209–4214.
33. Wang S, Zhang L, Zhang Z, Ding L, Zeng Q, Wang Z, Liang X, Gao M, Shen J, Xu H, Chen Q (2009). Photovoltaic effects in asymmetrically contacted CNT barrier-free bipolar diode. J Phys Chem C, 113(17): 6891–6893.
34. Zhou C, Wang S, Sun J, Wei N, Yang L, Zhang Z, Liao J, Peng LM (2013). Plasmonic enhancement of photocurrent in carbon nanotube by Au nanoparticles. Appl Phys Lett, 102 (10): 103102.
35. Huang H, Zhang D, Wei N, Wang S, Peng LM (2017). Plasmon-induced enhancement of infrared detection using a carbon nanotube diode. Adv Optical Mater, 5(6): 1600865.
36. Liang S, Ma Z, Wu G, Wei N, Huang L, Huang H, Liu H, Wang S, Peng LM (2016). Microcavity-integrated carbon nanotube photodetectors. ACS Nano, 10(7): 6963–6971.
37. Abdula D, Shim M (2008). Performance and photovoltaic response of polymer-doped carbon nanotube p–n diodes. ACS Nano, 2(10): 2154–2159.
38. Chen C, Song C, Yang J, Chen D, Zhu W, Liao C, Dong X, Liu X, Wei L, Hu N, He R (2017). Intramolecular pin junction photovoltaic device based on selectively doped carbon nanotubes. Nano Energy, 32: 280–286.
39. Chen C, Lu Y, Kong ES, Zhang YF, Lee ST (2008). Nanowelded carbon-nanotube-based solar microcells. Small, 4(9): 1313–1318.
40. Huang H, Wang F, Liu Y, Wang S, Peng LM (2017). Plasmonic enhanced performance of an infrared detector based on carbon nanotube films. ACS Appl Mater Interfaces, 9(14): 12743–12749.
41. Zhou H, Wang J, Ji C, Liu X, Han J, Yang M, Gou J, Xu J, Jiang Y (2019). Polarimetric vis-NIR photodetector based on self-aligned single-walled carbon nanotubes. Carbon, 143: 844–850.
42. Lu R, Li Z, Xu G, Wu JZ (2009). Suspending single-wall carbon nanotube thin film infrared bolometers on microchannels. Appl Phys Lett ,94(16): 163110.
43. Fernandes GE, Ho Kim J, Chin M, Dhar N, Xu J (2014). Carbon nanotube microbolometers on suspended silicon nitride via vertical fabrication procedure. Appl Phys Lett, 104(20): 201115.
44. Liu Y, Yin J, Wang P, Hu Q, Wang Y, Xie Y, Zhao Z, Dong Z, Zhu JL, Chu W, Yang N, Wei J, Ma W, Sun JL (2018). High-performance, ultra-broadband, ultraviolet to terahertz photodetectors based on suspended carbon nanotube films. ACS Appl Mater Interfaces, 10 (42): 36304–36311.

45. Nanot S, Cummings AW, Pint CL, Ikeuchi A, Akiho T, Sueoka K, Hauge RH, Léonard F, Kono J (2013). Broadband, polarization-sensitive photodetector based on optically-thick films of macroscopically long, dense and aligned carbon nanotubes. Sci Rep, 3(1): 1–7.
46. He X, Wang X, Nanot S, Cong K, Jiang Q, Kane AA, Goldsmith JE, Hauge RH, Léonard F, Kono J (2013) Photothermoelectric p–n junction photodetector with intrinsic broadband polarimetry based on macroscopic carbon nanotube films. ACS Nano 7(8): 7271–7277.
47. Straus DA, Tzolov M, Kuo TF, Yin A, Cardimona DA, Xu JM (2006). The carbon nanotube-silicon heterojunction as infrared detector. Photonics for Space Environments XI. International Society for Optics and Photonics, 6308: 63080Q.
48. Jia Y, Wei J, Wang K, Cao A, Shu Q, Gui X, Zhu Y, Zhuang D, Zhang G, Ma B, Wang L (2008). Nanotube-silicon heterojunction solar cells. Adv Mater, 20(23): 4594–4598.
49. Tune DD, Flavel BS (2018). Advances in carbon nanotube–silicon heterojunction solar cells. Adv Energy Mater, 8(15): 1703241.
50. An Y, Rao H, Bosman G, Ural A (2012). Characterization of carbon nanotube film-silicon Schottky barrier photodetectors. J Vacuum Sci Tech B, 30(2): 021805.
51. Scagliotti M, Salvato M, De Crescenzi M, Boscardin M, Castrucci P (2018). Influence of the contact geometry on single-walled carbon nanotube/Si photodetector response. Appl Nanosci, 8(5): 1053–1058.
52. Salvato M, Scagliotti M, De Crescenzi M, Boscardin M, Attanasio C, Avallone G, Cirillo C, Prosposito P, De Matteis F, Messi R, Castrucci P (2019). Time response in carbon nanotube/Si based photodetectors. Sens Actuator A, 292: 71–76.
53. Salvato M, Scagliotti M, De Crescenzi M, Crivellari M, Prosposito P, Cacciotti I, Castrucci P (2017). Single walled carbon nanotube/Si heterojunctions for high responsivity photodetectors. Nanotechnology, 28(43): 435201.
54. Riaz A, Alam A, Selvasundaram PB, Dehm S, Hennrich F, Kappes MM, Krupke R (2019). Near-infrared photoresponse of waveguide-integrated carbon nanotube-silicon junctions. Adv Electron Mater, 5(1): 1800265.
55. Kim YL, Jung HY, Park S, Li B, Liu F, Hao J, Kwon YK, Jung YJ, Kar S (2014). Voltage-switchable photocurrents in single-walled carbon nanotube–silicon junctions for analog and digital optoelectronics. Nature Photonics, 8: 239–243.
56. Chen HL, Cattoni A, De Lépinau R, Walker AW, Höhn O, Lackner D, Siefer G, Faustini M, Vandamme N, Goffard J, Behaghel B, Dupuis C, Bardou N, Dimroth F, Collin S (2019). A 19.9%-efficient ultrathin solar cell based on a 205-nm-thick GaAs absorber and a silver nanostructured back mirror. Nature Energy, 4: 761–767.
57. Liang C W, Roth S (2008). Electrical and optical transport of GaAs/carbon nanotube heterojunctions. Nano Lett, 8(7): 1809–1812.
58. Li H, Loke WK, Zhang Q, Yoon SF (2010). Physical device modeling of carbon nanotube/GaAs photovoltaic cells. Appl Phys Lett, 96: 043501.
59. Huo TT, Yin H, Zhou DY, Sun LJ, Tian T, Wei H, Hu NT, Yang Z, Zhang YF, Su YJ (2020). A self-powered broadband photodetector based on single-walled carbon nanotube/GaAs heterojunctions. ACS Sustainable Chem Eng, 8(41): 15532–15539.
60. Li G, Suja M, Chen M, Bekyarova E, Haddon RC, Liu J, Itkis ME (2017). Visible blind UV photodetector based on single-walled carbon nanotube thin film-ZnO vertical heterostructure. ACS Appl Mater Interfaces, 9: 37094–37104.
61. Yang M, Zhu JL, Liu W, Sun JL (2011). Novel photodetectors based on double-walled carbon nanotube film/TiO_2 nanotube array heterodimensional contacts. Nano Res, 4(9): 901–907.
62. Chuang CHM, Brown PR, Bulović V, Bawendi MG (2014). Improved performance and stability in quantum dot solar cells through band alignment engineering. Nature Mater, 13: 796–801.
63. Yang ZY, Fan JZ, Proppe AH, de Arquer FPG, Rossouw D, Voznyy O, Lan XZ, Liu M, Walters G, Quintero-Bermudez R, Sun B, Hoogland S, Botton GA, Kelley SO, Sargent EH (2017). Mixed-quantum-dot solar cells. Nature Commun, 8: 1325.

64. Yang XK, Yang J, Khan J, Deng H, Yuan SJ, Zhang J, Xia Y, Deng F, Zhou X, Umar F, Jin ZX, Song HS, Cheng C, Sabry M, Tang J (2020). Hydroiodic acid additive enhanced the performance and stability of PbS-QDs solar cells via suppressing hydroxyl ligand. Nano-Micro Lett, 12: 37.
65. Biswas C, Jeong H, Jeong MS, Yu WJ, Pribat D, Lee YH (2013). Quantum dot–carbon nanotube hybrid phototransistor with an enhanced optical stark effect. Adv Funct Mater, 23 (29): 3653–3660.
66. Gao L, Dong D, He J, Qiao K, Cao F, Li M, Liu H, Cheng YB, Tang J, Song HS (2014). Wearable and sensitive heart-rate detectors based on PbS quantum dot and multiwalled carbon nanotube blend film. Appl Phys Lett, 105(15): 153702.
67. Ka I, Le Borgne V, Ma D, El Khakani MA (2012). Pulsed laser ablation based direct synthesis of single-wall carbon nanotube/PbS quantum dot nanohybrids exhibiting strong, spectrally wide and fast photoresponse. Adv Mater, 24: 6289–6294.
68. Ka I, Le Borgne V, Fujisawa K, Hayashi T, Kim YA, Endo M, DL Ma, El Khakani MA (2016). Multiple exciton generation induced enhancement of the photoresponse of pulsed-laser-ablation synthesized single-wall-carbon-nanotube/PbS-quantum-dots nanohybrids. Sci Rep, 6(1): 1–11.
69. Tang Y, Fang H, Long M, Chen G (2018). Significant enhancement of single-walled carbon nanotube based infrared photodetector using PbS quantum dots. IEEE J Sel Top Quant, 24: 1–8.
70. Fujisawa K, Ka I, Le Borgne V, Kang CS, El Khakani MA (2016). Elucidating the local interfacial structure of highly photoresponsive carbon nanotubes/PbS-QDs based hanohybrids grown by pulsed laser deposition. Carbon, 96: 145–152.
71. Ka I, Le Borgne V, Fujisawa K, Hayashi T, Kim YA, Endo M, Ma DL. El Khakania MA (2020). PbS-quantum-dots/double-wall-carbon-nanotubes nanohybrid based photodetectors with extremely fast response and high responsivity. Mater Today Energy, 16: 100378.
72. Zheng J, Luo C, Shabbir B, Wang C, Mao W, Zhang Y, Huang Y, Dong Y, Jasieniak JJ, Pan C, Bao Q (2019). Flexible photodetectors based on reticulated SWNT/perovskite quantum dot heterostructures with ultrahigh durability. Nanoscale, 11(16): 8020–8026.
73. Spina M, Nafradi B, Tóháti H M, Kamarás K, Bonvin E, Gaal R, Forró L, Horvátha E (2016). Ultrasensitive 1D field-effect phototransistors: $CH_3NH_3PbI_3$ nanowire sensitized individual carbon nanotubes. Nanoscale, 8(9): 4888–4893.
74. Sundaram RS, Engel M, Lombardo A, Krupke R, Ferrari AC, Avouris Ph, Steiner M (2013). Electroluminescence in single layer MoS_2. Nano Lett, 13(4): 1416–1421.
75. Yang Z, Hong H, Liu F, Liu Y, Su M, Huang H, Liu K, Liang X, Yu WJ, Q. Vu A, Liu X, Liao L (2019). High-performance photoinduced memory with ultrafast charge transfer based on MoS_2/SWCNTs network van der Waals heterostructure, Small, 15: 1804661.
76. Zhang Z, Huang W, Xie Z, Hu W, Peng P, Huang G. (2017). Noncovalent functionalization of monolayer MoS_2 with carbon nanotubes: tuning electronic structure and photocatalytic activity. J Phys Chem C, 121: 21921–21929.
77. Jariwala D, Sangwan VK, Wu CC, Prabhumirashi PL, Geier ML, Marks TJ, Lauhon LJ, Hersam MC (2013). Gate-tunable carbon nanotube–MoS_2 heterojunction pn diode. Proc Nat Acad Sci, 110: 18076–18080.
78. Nguyen VT, Yim W, Park SJ, Son BH, Kim YC, Cao TT, Sim Y, Moon YJ, Nguyen VC, Seong MJ, Kim SK, Ahn YH, Lee S, Park JY (2018). Phototransistors with negative or ambipolar photoresponse based on as-grown heterostructures of single-walled carbon nanotube and MoS_2. Adv Funct Mater, 28: 1802572.
79. Furchi M, Urich A, Pospischil A, Lilley G, Unterrainer K, Detz H, Klang P, Andrews AM, Schrenk W, Strasser G, Mueller T (2012). Microcavity-integrated graphene photodetector. Nano Lett, 12: 2773–2777.
80. Liu Y, Cheng R, Liao L, Zhou HL, Bai JW, Liu G, Liu LX, Huang Y, Duan XF (2011). Plasmon resonance enhanced multicolour photodetection by graphene. Nature Commun, 2: 579.

81. Gan XT, Shiue RJ, Gao YD, Meric I, Heinz TF, Shepard K, Hone J, Assefa S, Englund D (2013). Chip-integrated ultrafast graphene photodetector with high responsivity. Nature Photonics, 7: 883–887.
82. Wang XM, Cheng ZZ, Xu K, Tsang HK, Xu JB (2013). High-responsivity graphene/silicon-heterostructure waveguide photodetectors. Nature Photonics, 7: 888–891.
83. Kim CO, Kim S, Shin DH, Kang SS, Kim JM, Jang CW, Joo SS, Lee JS, Kim JH, Choi SH, Hwang E (2014). High photoresponsivity in an all-graphene p–n vertical junction photodetector. Nature Commun, 5: 3249.
84. Liu CH, Chang YC, Norris TB, Zhong ZH (2014). Graphene photodetectors with ultra-broadband and high responsivity at room temperature. Nature Nanotechnol, 9: 273–278.
85. Zhang YZ, Liu T, Meng B, Li XH, Liang GZ, Hu XN, Wang QJ (2013). Broadband high photoresponse from pure monolayer graphene photodetector. Nature Commun, 4: 1811.
86. Chen ZF, Cheng ZZ, Wang JQ, Wan X, Shu C, Tsang HK, Ho HP, Xu JB (2015). High responsivity, broadband, and fast graphene/silicon photodetector in photoconductor mode. Adv Optical Mater, 3: 1207–1214.
87. Yu T, Wang F, Xu Y, Ma LL, Pi XD, Yang DR (2016). Graphene coupled with silicon quantum dots for high-performance bulk-silicon-based Schottky-junction photodetectors. Adv Mater, 28(24): 4912–4919.
88. Wu Y, Yan X, Zhang X, Ren X (2016). A monolayer graphene/GaAs nanowire array Schottky junction self-powered photodetector. Appl Phys Lett, 109: 183101.
89. Luo LB, Chen JJ, Wang MZ, Hu H, Wu CY, Li Q, Wang L, Huang JA, Liang FX (2014). Near-infrared light photovoltaic detector based on GaAs nanocone array/monolayer graphene Schottky junction, Adv Funct Mater, 24: 2790–2800.
90. Tao ZJ, Zhou DY, Yin H, Cai BF, Huo TT, Ma J, Di ZF, Hu NT, Yang Z, Su YJ (2020). Graphene/GaAs heterojunction for highly sensitive, self-powered visible/NIR photodetectors. Mater Sci Semiconductor Proc, 111: 104989.
91. Zeng LH, Wang MZ, Hu H, Nie B, Yu YQ, Wu CY, Wang L, Hu JG, Xie C, Liang FX, Luo LB (2013). Monolayer graphene/germanium Schottky junction as high-performance self-driven infrared light photodetector. ACS Appl Mater Interfaces, 5(19): 9362–9366.
92. Kong WY, Wu GA, Wang KY, Zhang TF, Zou YF, Wang DD, Luo LB (2016). Graphene-β-Ga_2O_3 heterojunction for highly sensitive deep UV photodetector application. Adv Mater, 28(48): 10725–10731.
93. Dang VQ, Trung TQ, Kim DI, Duy LT, Hwang BU, Lee DW, Kim BY, Toan LD, Lee NE (2015). Ultrahigh responsivity in graphene-ZnO nanorod hybrid UV photodetector. Small, 11(25): 3054–3065.
94. Nie B, Hu JG, Luo LB, Xie C, Zeng LH, Lv P, Li FZ, Jie JS, Feng M, Wu CY, Yu YQ, Yu SH (2013). Monolayer graphene film on ZnO nanorod array for high-performance Schottky junction ultraviolet photodetectors. Small, 9(17): 2872–2879.
95. Konstantatos G, Howard I, Fischer A, Hoogland S, Clifford J, Klem E, Levina L, Sargent EH (2006). Ultrasensitive solution-cast quantum dot photodetectors. Nature, 442(7099): 180–183.
96. Sukhovatkin V, Hinds S, Brzozowski Lu, Sargent EH (2009). Colloidal quantum-dot photodetectors exploiting multiexciton generation. Science, 324(5934): 1542–1544.
97. Clifford JP, Konstantatos G, Johnston K W, Hoogland S, Levina L, Sargent EH (2009). Fast, sensitive and spectrally tuneable colloidal-quantum-dot photodetectors. Nature Nanotechnol, 4(1): 40–44.
98. Su Y, Lu X, Xie M, Geng HJ, Wei H, Yang Z, Zhang YF (2013). A one-pot synthesis of reduced graphene oxide-Cu_2S quantum dot hybrids for optoelectronic devices. Nanoscale, 5: 8889–8893.
99. Jiang G, Su YJ, Li M, Hu J, Zhao B, Yang Z, Wei H (2016). Synthesis and optoelectronic properties of reduced graphene oxide/InP quantum dot hybrids. RSC Adv, 6: 97861–97864.

100. Konstantatos G, Badioli M, Gaudreau L, Osmond J, Bernechea M, de Arquer FPG, Gatti F, Koppens FHL (2012). Hybrid graphene–quantum dot phototransistors with ultrahigh gain. Nature Nanotechol, 7: 363–368.
101. Zhang DY, Gan L, Cao Y, Wang Q, Qi LM, Guo XF (2012). Understanding charge transfer at PbS-decorated graphene surfaces toward a tunable photosensor. Adv Mater, 24: 2715–2720.
102. Che Y, Zhang Y, Cao X, Zhang H, Song X, Cao M, Yu Y, Dai H, Yang J, Zhang G, Yao J (2017). Ambipolar graphene–quantum dot hybrid vertical photodetector with a graphene electrode. ACS Appl Mater Interfaces, 9(37): 32001–32007.
103. Nian Q, Gao L, Hu YW, Deng BW, Tang J, Cheng GJ (2017). Graphene/PbS-quantum dots/graphene sandwich structures enabled by laser shock imprinting for high performance photodetectors. ACS Appl Mater Interfaces, 9(51): 44715–44723.
104. Kramer NJ, Schramke KS, Kortshagen UR (2015). Plasmonic properties of silicon nanocrystals doped with boron and phosphorus. Nano Lett, 15: 5597–5603.
105. Zhou S, Pi XD; Ni ZY, D Y, Jiang YY, Jin CH, Delerue C, Yang DR, Nozaki T (2015). Comparative study on the localized surface plasmon resonance of boron- and phosphorus-doped silicon nanocrystals. ACS Nano, 9(1): 378–386.
106. Ni Z, Ma L, Du S, Xu Y, Yuan M, Fang H, Wang Z, Xu M, Li D, Yang J, Hu W, Pi X, Yang D (2017). Plasmonic silicon quantum dots enabled high-sensitivity ultrabroadband photodetection of graphene-based hybrid phototransistors. ACS Nano, 11(10): 9854–9862.
107. Hu AQ, Tian HJ, Liu QL, Wang L, Wang L, He XY, Luo Y, Guo X (2019). Graphene on self-assembled InGaN quantum dots enabling ultrahighly sensitive photodetectors. Adv Optical Mater, 7(8): 1801792.
108. Sun T, Wang Y, Yu W, Wang Y, Dai Z, Liu Z, Shivananju B, Zhang Y, Fu K, Shabbir B, Ma W, Li S, Bao Q (2017). Flexible broadband graphene photodetectors enhanced by plasmonic $Cu_{3-x}P$ colloidal nanocrystals. Small, 13(42): 1701881.
109. Gong MG, Sakidja R, Liu QF, Goul R, Ewing D, Casper M, Stramel A, Elliot A, Wu JZ (2018). Broadband photodetectors enabled by localized surface plasmonic resonance in doped iron pyrite nanocrystals. Adv Optical Mater, 6(8): 1701241.
110. Pan R, Li HY, Wang J, Jin X, Li QH, Wu ZM, Gou J, Jiang YD, Song YL (2018). High-responsivity photodetectors based on formamidinium lead halide perovskite quantum dot-graphene hybrid. Part Part Syst Charact, 35(4): 1700304.
111. Kwak DH, Lim DH, Ra HS, Ramasamy P, Lee JS (2016). High performance hybrid graphene-$CsPbBr_{3-x}I_x$ perovskite nanocrystal photodetector. RSC Adv, 6: 65252–65256.
112. Dean CR, Young AF, Meric I, Lee C, Wang L, Sorgenfrei S, Watanabe K, Taniguchi T, Kim P, Shepard KL, Hone J (2010). Boron nitride substrates for high-quality graphene electronics. Nature Nanotechnol, 5: 722–726.
113. Zhang W, Chuu CP, Huang JK, Chen CH, Tsai ML, Chang YH, Liang CT, Chen YZ, Chueh YL, He JH, Chou MY, Li LJ (2014). Ultrahigh-gain photodetectors based on atomically thin graphene-MoS_2 heterostructures. Sci Rep, 4: 3826.
114. Li X, Lin S, Lin X, Xu Z, Wang P, Zhang S, Zhong H, Xu W, Wu Z, Fang W (2016). Graphene/h-BN/GaAs sandwich diode as solar cell and photodetector. Opt Express, 24(1): 134–145.
115. Li H, Li X, Park JH, Tao L, Kim KK, Lee YH, Xu JB (2019). Restoring the photovoltaic effect in graphene-based van der Waals heterojunctions towards self-powered high-detectivity photodetectors. Nano Energy, 57: 214–221.
116. Kang B, Kim Y, Yoo WJ, Lee C (2018). Ultrahigh photoresponsive device based on ReS_2/graphene heterostructure. Small, 14(45): 1802593.
117. Yu WJ, Liu Y, Zhou H, Yin A, Li Z, Huang Y, Duan XF (2013). Highly efficient gate-tunable photocurrent generation in vertical heterostructures of layered materials. Nature Nanotechnol, 8(12): 952–958.
118. Britnell L, Ribeiro RM, Eckmann A, Jalil R, Belle BD, Mishchenko A, Kim YJ, Gorbachev RV, Georgiou T, Morozov SV, Grigorenko AN (2013). Strong light-matter interactions in heterostructures of atomically thin films. Science, 340: 1311–1314.

119. Massicotte M, Schmidt P, Vialla F, Schädler KG, Reserbat-Plantey A, Watanabe K, Taniguchi T, Tielrooij KJ, Koppens FH (2016). Picosecond photoresponse in van der Waals heterostructures. Nature Nanotechnol ,11: 42–46.
120. Long M, Liu E, Wang P, Gao A, Xia H, Luo W, Wang B, Zeng J, Fu Y, Xu K, Zhou W (2016). Broadband photovoltaic detectors based on an atomically thin heterostructure. Nano Lett, 16: 2254–2259.
121. Li A, Chen Q, Wang P, Gan Y, Qi T, Wang P, Tang F, Wu JZ, Chen R, Zhang L, Gong Y (2019). Ultrahigh-sensitive broadband photodetectors based on dielectric shielded MoTe/graphene/SnS p-g-n junctions. Adv Mater, 31(6): 1805656–1805664.
122. Lu R, Liu J, Luo H, Chikan V, Wu JZ (2016). Graphene/GaSe-nanosheet hybrid: towards high gain and fast photoresponse. Sci Rep, 6: 19161.
123. Chen Z, Biscaras J, Shukla A (2015). A high performance graphene/few-layer InSe photodetector. Nanoscale, 7: 5981–5986.
124. Luo W, Cao Y, Hu P, Cai K, Feng Q, Yan F, Yan T, Zhang X, Wang K (2015). Gate tuning of high-performance InSe-based photodetectors using graphene electrodes. Adv Optical Mater, 3: 1418–1423.
125. Mudd GW, Svatek SA, Hague L, Makarovsky O, Kudrynskyi ZR, Mellor CJ, Beton PH, Eaves L, Novoselov KS, Kovalyuk ZD, Vdovin EE (2015). High broad-band photoresponsivity of mechanically formed InSe-graphene van der Waals heterostructures. Adv Mater, 27 (25): 3760–3766.
126. Kim J, Park S, Jang H, Koirala N, Lee JB, Kim UJ, Lee HS, Roh YG, Lee H, Sim S, Cha S (2017). Highly Sensitive, gate-tunable, room-temperature mid-infrared photodetection based on graphene-Bi_2Se_3 heterostructure. ACS Photonics, 4: 482–488.
127. Qiao H, Yuan J, Xu Z, Chen C, Lin S, Wang Y, Song J, Liu Y, Khan Q, Hoh HY, Pan CX (2015), Broadband photodetectors based on graphene-Bi_2T_{e3} heterostructure. ACS Nano. 9 (2): 1886–1894.
128. Li L, Kim J, Jin C, Ye GJ, Qiu DY, Felipe H, Shi Z, Chen L, Zhang Z, Yang F, Watanabe K (2017). Direct observation of the layer-dependent electronic structure in phosphorene. Nature Nanotechnol, 12(1): 21–25.
129. Cai Y, Zhang G, Zhang YW (2014). Layer-dependent band alignment and work function of few-layer phosphorene. Sci Rep, 4: 6677.
130. Deng Y, Luo Z, Conrad NJ, Liu H, Gong Y, Najmaei S, Ajayan PM, Lou J, Xu X, Ye PD (2014). Black phosphorus-monolayer MoS_2 van der Waals heterojunction p–n diode. ACS Nano, 8(8): 8292–8299.
131. Huang M, Li S, Zhang Z, Xiong X, Li X, Wu Y (2017). Multifunctional high-performance van der Waals heterostructures. Nature Nanotechnol, 12: 1148–1154.
132. Liu Y, Shivananju BN, Wang Y, Zhang Y, Yu W, Xiao S, Sun T, Ma W, Mu H, Lin S, Zhang H (2017). Highly efficient and air stable infrared photodetector based on 2D layered graphene-black phosphorus heterostructure. ACS Appl Mater Interfaces, 9(41): 36137–36145.
133. Zhou GG, Li ZJ, Ge YQ, Zhang H, Sun ZH (2020). A self-encapsulated broadband phototransistor based on a hybrid of graphene and black phosphorus nanosheets. Nanoscale Adv, 2: 1059–1065.
134. Lu Q, Yu L, Liu Y, Zhang J, Han G, Hao Y (2019). Low-noise mid-infrared photodetection in BP/h-BN/Graphene van der Waals heterojunctions. Mater, 12(16): 2532.

Chapter 6
All-Carbon van der Waals Heterojunction Photodetectors

6.1 Introduction

The emergence of van der Waals (vdW) heterostructures composed of distinct two-dimensional (2D) layered materials offers almost unlimited possibilities for novel nanodevices and related structure designs, [1–4] in which the typical lattice matching is no longer needed and the physical properties of heterojunctions is determined by both each 2D materials and the electron coupling effect at the interface. In addition, the atomically sharp interface contact can effectively avoid the existence of disordered atomic transition layers, which are usually the trap sites of photogenerated carriers. Actually, the low-dimensional materials with passivated and dangling-bond-free surface can also be used to prepare the heterostructures through vdW force interactions, forming a large number of mixed-dimensional vdW heterostructures [4]. As a consequence, more novel optoelectronic nanodevices will be designed by creating the vdW heterostructures with varying bandgap alignments using different mixed-dimensional nanomaterials.

As we know, the low-dimensional carbon allotropes can behave as bandgap-tunable semiconductors or high-mobility conductors according to their atomic arrangements. For example, zero-dimensional (0D) fullerenes are typical n-type semiconductors, which is widely used as electron transporting materials. Another 0D nanocarbon (carbon-based quantum dots (QDs)) has shown novel optical and electric properties, such as high light-absorption ability, fluorescence upconversion characteristics, and so on [5–7]. As a quasi-one-dimensional (1D) carbon nanostructure, single-walled carbon nanotubes (SWCNTs) exhibit highly chirality-depended physical and chemical properties. One third of SWCNTs behave metallic with ultrahigh electron mobility as high as 10^5 cm^2 V^{-1} s^{-1} at room temperature, and the remaining two-thirds are weak p-type direct-bandgap semiconductors in air with a bandgap inversely proportional to their diameter. Furthermore, the wide light absorption (UV-near infrared) and high absorption coefficient (10^4–10^5 cm^{-1}) enable sc-SWCNTs with only an ultrathin thickness to

Y. Su, *High-Performance Carbon-Based Optoelectronic Nanodevices*,
Springer Series in Materials Science 319,
https://doi.org/10.1007/978-981-16-5497-8_6

build high performance photodetectors [8, 9]. For two-dimensional (2D) graphene, the ultrahigh mobility exceed 10^6 cm^2 V^{-1} s^{-1} makes it to be an ideal carrier transporting channel in graphene-based vdW heterojunctions. Meanwhile, unique electronic structure of graphene enables the heterojunction to exhibit the unusual electronic and optoelectronic properties after interfacing with other low-dimensional materials. Therefore, the abovementioned excellent electric and optical properties of carbon nanomaterials inspire the design of novel all-carbon vdW heterostructures (such as 0D/1D, 0D/2D and 1D/2D heterostructures), which can be assembled by simply integrating the carbon allotropes with different optical and electric properties. Especially, the contact type and area in these all-carbon vdW heterostructures is very different from those in typical 2D/2D vdW heterostructures, exceptional interface electron coupling interaction or charge transfer dynamic may be expected in the carbon-based photodetectors.

In this chapter, we focus on the all-carbon vdW heterojunction photodetectors based on mixed-dimensional carbon nanomaterials, namely C_{60}/SWCNT (0D/1D), C_{60}/graphene (0D/2D), carbon QD/graphene (0D/2D) and SWCNT/graphene (1D/2D). Different stacking methods of all-carbon vdW heterostructures is first presented, and then we summarize systemically the recent research progress on all-carbon vdW heterojunction photodetectors in turn. Finally, we discuss the current challenges and highlight the future perspectives of all-carbon vdW heterojunction photodetectors.

6.2 The Stacking Methods of All-Carbon vdW Heterostructures

The controlled manual-stacking of high quality all-carbon vdW heterostructures is the basis of building carbon-based photodetectors [10]. Different approaches have been reported in the previous works, but the main principle is to keep the interface clean and the thickness or network controllable. Especially, the electron coupling effect at the interface is expected to be influenced easily in the all-carbon vdW heterojunctions due to similar atomic hybridization. Ultraclean interface and defect-free active materials are highly desired for high-performance all-carbon vdW heterojunction photodetectors. For the vdW heterostructures composed of fullerene and SWCNT (or graphene), the thermal evaporation of fullerene has been usually used to form high quality and clean interfaces, which is very critical for highly effective charge transfer at the interfaces. When fullerene is replaced by fullerene derivatives, the spin-coating is usually used to form uniform thin film. Similarly, the dip-coating and spin-coating methods have been commonly adopted to prepare carbon QD film, because carbon QDs with rich chemical functional groups are prepared using solution method.

For CNT/graphene vdW heterostructures, the formation of these 1D/2D vdW heterostructures can be carried out by direct growth and transfer process.

Especially, direct growth method is good choice for the formation of high quality SWCNT-graphene vdW heterostructures. Usually, the two-step CVD growth method has been adopted, in which SWCNT (or graphene) was synthesized firstly using catalysts and graphene (or SWCNT) was grown subsequently on the surface of SWCNT (or graphene). For instance, Zhu et al. [11] firstly reported a covalently bonded SWCNT/graphene vdW heterostructures by growing vertically aligned SWCNTs seamlessly on the surface of graphene. In this 3D vdW heterostructure, the covalent transformation of sp^2 carbon between the SWCNTs and graphene at the atomic resolution level had been demonstrated, resulting in an ohmic contact between SWCNT and graphene. Ultrathin planar SWCNT/graphene films have also been synthesized directly by chemical vapor deposition [12, 13]. Shi et al. [14] demonstrated a SWCNT network/graphene vdW heterostructure films with high conductivity and mechanical flexibility via the CVD growth of graphene using SWCNT networks on Cu surface as a porous template, as shown in Fig. 6.1. Furthermore, monolithic all-carbon transistors could be fabricated directly by chemical synthesis, in which graphene as active elements and SWCNT/graphene hybrids as electrodes. Similar ultrathin SWCNT/graphene films have also been reported by direct growth monolayer graphene on the surface of random SWCNT networks [15]. The vdW heterostructure film show enhanced mechanical strength without degrading their high optical transmittance. Interesting, the interactions between SWCNT and graphene can induce a strong photoresponse in the region where SWCNTs link different graphene domains together. In addition, the SWCNT/graphene vdW heterostructures can be also synthesized by the growth of graphene from the edges of a partially unzipped SWCNT [16, 17].

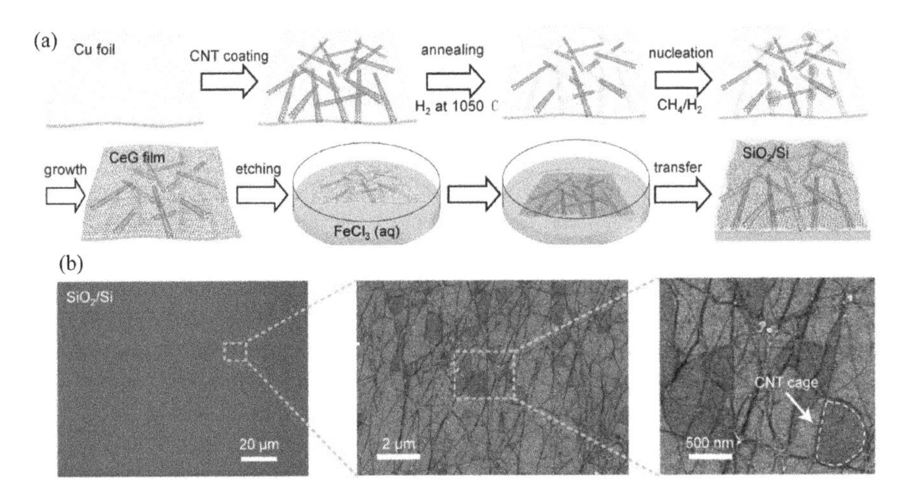

Fig. 6.1 **a** Schematic illustration of the growth process of SWCNT/graphene vdW heterostructures. **b** Optical and SEM images of the all-carbon vdW heterostructure film on SiO_2/Si substrate. Reproduced with permission [14]. Copyright 2014, WILEY–VCH

Except for direct synthesis, the wet transfer process is usually been used to prepare the SWCNT/graphene vdW heterostructure, in which individual SWCNT networks (or graphene) or both of them was transferred onto the target substrates. Usually, the SWCNT/graphene vdW heterostructures are prepared by transferring graphene nanosheets on the surface of SWCNT networks with poly(methyl methacrylate) (PMMA) as a transferring medium [18, 19]. Sometimes, the SWCNT networks are also transferred on the surface of graphene to form all-carbon vdW heterostructures [20]. Both SWCNT network and graphene nanosheets have been transferred [21]. In addition, the SWCNT/graphene vdW heterostructure film has also been prepared by spray coating of SWCNTs on the surface of CVD-graphene [22].

6.3 SWCNT/C_{60} vdW Heterostructures

As we know, the sc-SWCNTs exhibit a weak p-type behavior with ultrahigh carrier mobility in air, while fullerenes (e.g., C_{60}, C_{70}) is an excellent n-type semiconductor as electron transporting layer with the long exciton diffusion length (up to 40 nm). When the sc-SWCNTs combine with the C_{60} to form all-carbon p-n junctions, the photogenerated excitons in sc-SWCNTs are dissociated into free electron-hole pairs due to the efficient electron transfer to C_{60} at the interface. More importantly, the photoinduced electron transfer time is as fast as 120 fs and the yield of charge transfer is $\sim$38 $\pm$ 3%, [23] which enables the sc-SWCNT/C_{60} vdW heterostructures to be the great potential for photovoltaic and photodetectors. In the past ten years, the photovoltaic devices have been extensively studied based on stacked SWCNT/C_{60} p/n junctions, [24–26] where sc-SWCNTs are used as light absorption materials and C_{60} as electron transportation layer is used to form suitable build-in electric field at the interface to dissociate excitons most efficiently generated in the SWCNT layers. But the SWCNT/C_{60} hybrids are less used to build high sensitive photodetectors. Park et al. [27] demonstrated firstly a sc-SWCNT/C_{60} hybrid planar phototransistor (Fig. 6.2) with the C_{60} serving as electron traps and the sc-SWCNTs as an active light absorbing/carrier generating layer, in which high gain (holes per generated electron/hole pair) and fine-tuned photoresponse could be required using gate voltage. Under light illuminations, the photogenerated electrons are trapped by C_{60} while the holes photogenerated in the sc-SWCNTs are transported laterally from source to drain electrodes, which is contrary to the previous vertically stacked photodiodes where C_{60} and sc-SWCNTs are used to transport the photogenerated electrons and holes toward the corresponding collecting electrodes, respectively. The responsivity increases linearly as a function of source-drain voltage (V_{DS}) and a high responsivity nearly 200 A W^{-1} was obtained at V_{DS} of -1 V and gate voltage of 0 V. The max responsivity was 97.5 A W^{-1} at V_{DS} of 0 V and gate voltage of -2 V, and then the responsivity decreased due to the reduction of the electron number when the gate voltage was smaller than -2 V. Such a high responsivity of the device is correlated to the number of the carriers collected per

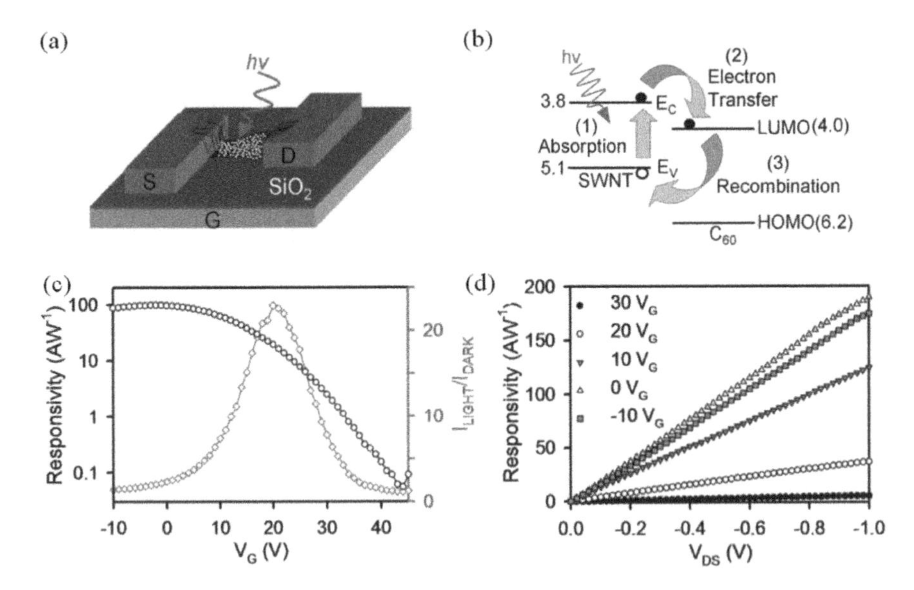

Fig. 6.2 **a** Schematic diagram of sc-SWCNT/C_{60} vdW heterostructure phototransistor. **b** Schematic of light absorption and subsequent electron transfer and recombination processes in the heterojunctions. **c** Responsivity, noise spectral density and **d** specific detectivity as a function of frequency. Reproduced with permission [27]. Copyright 2015, WILEY–VCH

photoinduced carrier, and the photoconductive gain (G) is estimated to be (2–4) $\times$ 10^4. Bergemann et al. [26] also developed sc-SWCNT/C_{60} heterojunction phototransistors by thermal evaporating C_{60} layer on aligned SWCNT arrays. Surprisingly, the device exhibited an ultrahigh responsivity of 10^8 A W^{-1} under UV or green illumination and 720 A W^{-1} toward 950 nm illumination. Such a high performance is mainly attributed to the thick C_{60} sensitizer layer and the strong build-in electric field at the C_{60}/SWCNT interface due to favorable band alignment.

6.4 Graghene/C_{60} vdW Heterostructures

The graphene/C_{60} heterostructures have also been demonstrated to fabricate photoelectric nanodevices by combining the excellent electrical and optical properties of graphene and C_{60}. For example, Qin et al. [28] demonstrated an all-carbon heterostructure phototransistor by assembling C_{60} on the surface of graphene. Under UV illumination, a high responsivity is evaluated to be $\sim 10^7$ A W^{-1} with the gate-tunable response time because of enhanced optical absorption and efficient exciton dissociation at the graphene/C_{60} interface. Moreover, the as-fabricated 5 $\times$ 5 photodetector array exhibited excellent environmental robustness (Fig. 6.3). Chugh et al. [29] demonstrated a multilayer graphene/C_{60} vdW heterostructure photodetectors using electrophoretic deposition. Compared with bare graphene, the

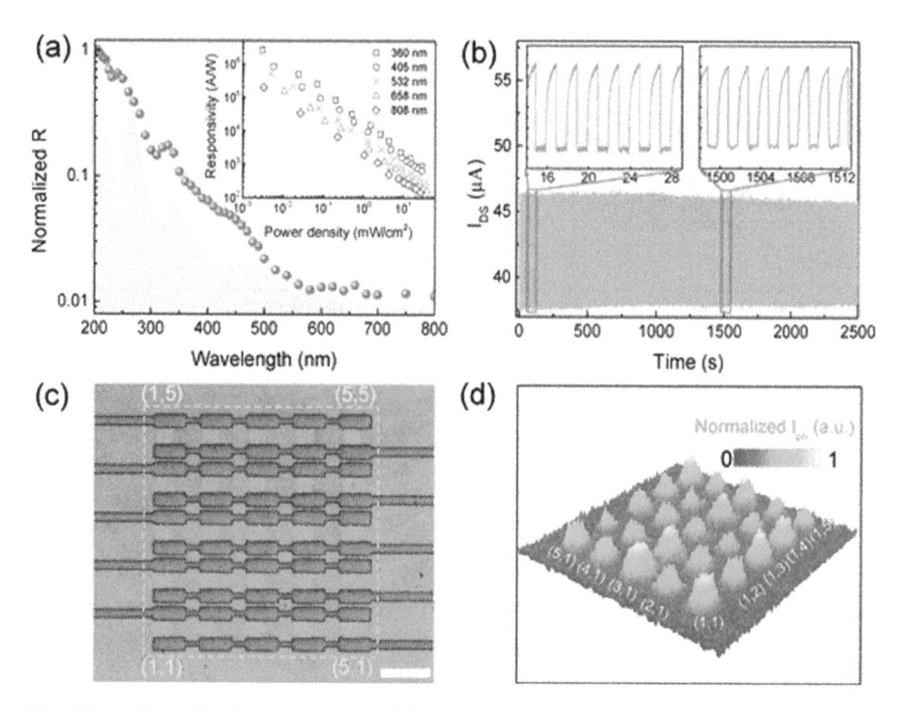

Fig. 6.3 **a** Normalized spectral responsivity of the hybrid device and corresponding to the absorption spectra of the hybrid film. **b** The reproducible photoresponse curves at ambient conditions. **c** Optical image and **d** the spatial-light mapping for the 5 × 5 arrayed devices. Reproduced with permission [28]. Copyright 2018, American Chemical Society

blue shifts and the reduction in I_{2D}/I_G intensity of Raman peaks (G- and 2D-bands) in the heterostructures mean that graphene was p-doped (or hole-doped) after forming heterostructures with C_{60}. When the devices were irradiated with a tunable laser of 400 nm, a strong photoresponse could be measured with ultrahigh responsivity of $\sim 10^9$ A W^{-1}, detectivity of $\sim 10^{15}$ Jones and EQE value of $\sim 10^9\%$. Such a high photoresponse performance is attributed to the doping enhancement in graphene due to the heterostructure formation with C_{60}.

Qin et al. [30] further fabricated a room-temperature all-carbon phototransistor based on planar graphene-C_{60}-graphene heterostructures with h-BN as molecular-wetting substrate, where well-crystallized C_{60} film as a photoconductive channel was deposited on and between graphene contact electrodes. The introduction of h-BN substrate was used to minimize the charge traps at the interfaces and achieve a well-ordered structure. Under light illumination, the gate-tunable all-carbon phototransistor exhibited a high responsivity of 5510 A W^{-1} (360 nm) and 2280 A W^{-1} (405 nm) with an operating bandwidth of ~ 6 kHz. Vertical graphene-C_{60}-graphene heterostructures have also been demonstrated to fabricate phototransistors [31]. These devices showed a high responsivity of 3.4×10^5 A W^{-1} (405 nm) and fast response speed (23 ms), and 250 × 250 vertical graphene/C_{60}/graphene phototransistors array was also fabricated. Interestingly, a bi-directional response photoresponse (positive and negative) has been observed due to different charge transfer mechanism. In addition, C_{60} has also been used to trap photogenerated carriers in the graphene nanoribbon-C_{60} heterostructures, and longer carrier recombination lifetimes can be achieved due to a high electron trapping efficiency of the C_{60} layer, resulting in a high gate-tunable responsivity of 0.4 A W^{-1} under MIR illumination at room temperature [32].

6.5 Graghene/Carbon QD vdW Heterostructures

Except for fullerenes, carbon-based QDs are another important 0D carbon nanomaterial, which can be divided simply into two types: sphere-like carbon QDs (CQDs) and layer-liked graphene QDs (GQDs). Both CQDs and GQDs have been attracting an increasing attention in the past few years due to their novel size- and functional group-tunable optical and chemical properties [33–35]. By combining the large absorptivity of carbon-based QDs with high carrier mobility of graphene, new type 0D/2D all-carbon vdW heterostructures are formed. For example, Cheng et al. [36] firstly demonstrated all-carbon active materials consisting of graphite QD/graphene for ultrahigh sensitive photodetector. Under 325 nm light illumination, the photogenerated electron/hole pairs in graphite QDs could be separated efficiently at the interface by the appropriate build-in electric field formed between graphite QDs and graphene, resulting in high responsivity of 4×10^7 A W^{-1}. Tetsuka et al. [37] developed a high-performance flexible nitrogen-functionalized GQD/graphene UV photodetector, exhibits high responsivity of ca. 1.5×10^4 A W^{-1} and detectivity of ca. 5.5×10^{11} Jones for deep-UV light as short as 255 nm. BN nanosheets as a buffer layer could be used to facilitate the photoexcited carrier separation at the interface between nitrogen-functionalized QDs and graphene, further improving the photodetector performance with a responsivity of ca. 2.3×10^6 A W^{-1} and detectivity of ca. 5.5×10^{13} Jones in the deep UV region [38]. Meanwhile, a high responsivity of 3.4×10^2 A W^{-1} was still obtained the near-infrared (NIR) region. By choosing the GQDs with different sizes and edge structures, the GQD/graphene vdW heterostructure photodetectors can be used to detect different light in the UV-vis range [39]. The device exhibits the high responsivity at the maximal specific absorption wavelengths of each GQD, and the highest responsivity was evaluated to be as high as 10^7 A W^{-1} under 400 nm illuminations. A comparison of the photoelectric performance of the photodetectors is displayed in Table 6.1.

In addition, the GQD/graphene photodetectors have also been demonstrated based on multiple-layer GQDs sandwiched between graphene sheets [40]. A responsivity of 0.2–0.5 A W^{-1} and detectivity of $>10^{11}$ Jones could be achieved in the broad spectral range from UV to NIR. Interestingly, Haider et al. [41]

Table 6.1 Comparison of the performance of photodetectors based on the GQD/graphene heterojunctions

Active material	Wavelength (nm)	Responsivity (A W^{-1})	Detectivity (Jones)	t_{rise}	t_{fall}	Ref.
GQD/GR	325	4×10^7	/	2 s	/	[36]
N-GQD/GR	255–405	$\sim 1.5 \times 10^4$	$\sim 5.5 \times 10^{11}$	3.7 s	22 s	[37]
N-GQD/BN/GR	254–940	$\sim 2.3 \times 10^6$	$\sim 5.5 \times 10^{13}$	45 ms	/	[38]
GQD/GR	UV-vis	$\sim 10^7$ at 400 nm	6×10^9	/	300 s	[39]
GQD/GR	300–1100	0.2–0.5	$>10^{11}$	~ 2 μs	~ 20 μs	[40]
GQD/GR/PZT	325	4.06×10^9	/	0.3 s	4.9 s/	[41]

demonstrated an ultrasensitive GQD/graphene vdW heterostructure photodetector on the $Pb(Zr_{0.2}Ti_{0.8})O_3$ (PZT) substrate over a wide range of illumination power, in which the intrinsic electric field from PZT substrate could separate the photogenerated electron/hole pairs in GQDs and the as-provided built-in electric field facilitated the hole transfer from GQDs to graphene. As a result, the photodetector exhibited an ultrahigh sensitivity as high as 4.06×10^9 A W^{-1} under 325 nm light illumination.

6.6 Graphene/SWCNTs vdW Heterostructures

Compared with the 0D/graphene and 0D/SWCNTs, the graphene/SWCNT vdW heterostructures as the active materials of photodetectors have more advantages as following: (1) the formation of graphene/SWCNT vdW heterostructures facilitates the efficient dissociation of robust excitons in the SWCNTs at the interface [42]; (2) ultrahigh carrier mobility of both graphene and SWCNTs facilitates the fast transportation of the photogenerated electrons and holes; (3) high lattice similarity facilitates tighter electron coupling due to similar sp^2 hybridization of carbon atoms, and the physical properties of all-carbon vdW heterostructures can be regulated by changing the electron coupling at the interface; (4) unique band structures of both sc-SWCNTs and graphene enable us to tailor their Fermi levels more flexible, thereby controlling the built-in field and the charge transfer dynamic at the interface. Therefore, these 1D/2D all-carbon vdW heterostructures as active materials for high performance photodetectors have attracted increasing attentions in the past few years.

6.6.1 Charge Transfer Between Graphene and SWCNTs

As we know, the strong Coulomb interaction in SWCNTs makes excitons have large binding energy, the efficient exciton dissociation at the interface is crucial to fabricate a high-performance photodetector. In the graphene/SWCNT vdW heterostructures, the efficient charge transfer at the interface is a key factor for high performance photoelectric devices, which can be monitored using the Raman shift and XPS peak shift. Rao et al. [43] demonstrated a static electron transfer from graphene to the SWCNTs in the graphene/SWCNT vdW heterostructures, resulting in p-type doped graphene and n-type doped SWCNTs. Liu et al. [44] investigated both static and dynamic charge transfer processes between SWCNT and graphene in the graphene/SWCNT vdW heterostructures, and demonstrated that graphene was found to be p-type doped with (6, 5) sc-SWCNTs, while n-type doped with m-SWCNTs. Moreover, the charge transfer at the interface can be further controlled by changing the local environment (moisture or substrate) or bias voltage [45, 46]. On the other hand, our experimental results also demonstrated that semiconductor (Ge or Si) as substrates modified the charge transfer between graphene and sc-SWCNTs due to the change of Fermi levels in different dopant types. These abovementioned results not only make the graphene/SWCNT vdW heterostructures one of the most promising candidates for the next-generation photodetectors, but also provide important design guidelines for the graphene/SWCNT vdW heterostructure based photoelectric devices.

6.6.2 Graphene/SWCNT vdW Heterojunction Photodetectors

By combing excellent electronics and photoelectric properties of SWCNTs and graphene, novel high-performance all-carbon vdW heterojunction photodetectors are expected to be fabricated by controlling the bandgap (diameter) of SWCNTs, the Fermi level and their interface. For example, Liu et al. [18] demonstrated an all-carbon hybrid broad phototransistor based on the graphene/atomically thin SWCNT vdW heterostructures (Fig. 6.4). Large built-in electric field at the interface not only promotes the effective separation of robust excitons in SWCNTs, but also reduces remarkably the recombination probability of spatially isolated photogenerated electrons and holes. The devices exhibited a broad photoresponse across visible to NIR range (400–1550 nm) when different SWCNTs were used. The max responsivity of >100 A W^{-1} and a fast response time of $\sim$100 μs have been achieved in the graphene/(6,5) SWCNT vdW heterostructure devices under 650 nm illuminations. This all-carbon vdW heterostructure phototransistors can also be fabricated by covering graphene film on the arc-discharge sc-SWCNT film [47]. Under 532 nm light illumination, the photoexcited electrons are transferred from sc-SWCNTs to graphene while the holes are trapped in the sc-SWNTs due to the

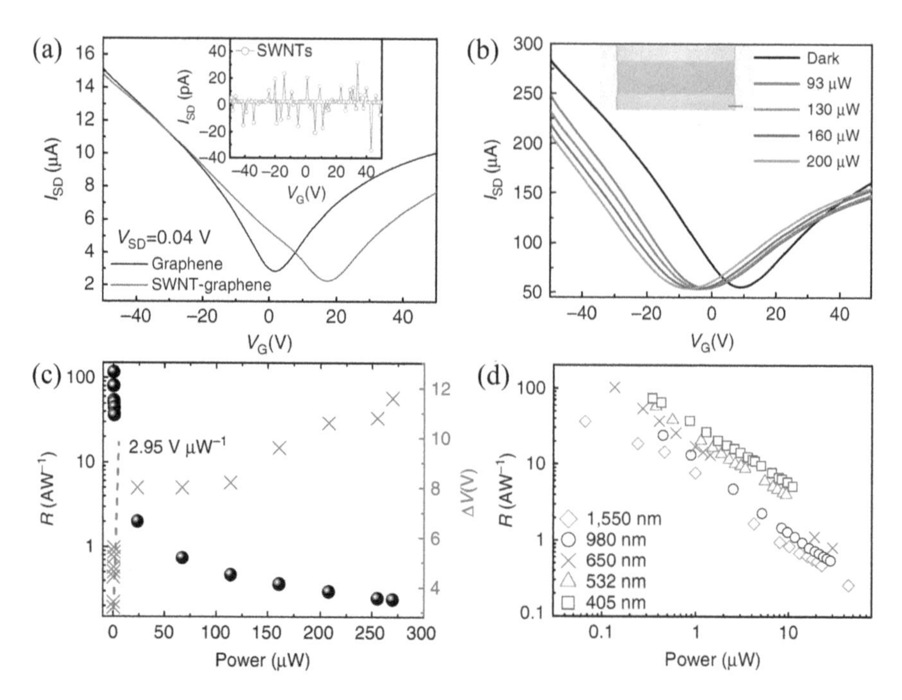

Fig. 6.4 Source-drain current curves of SWCNT/graphene vdW heterostructure transistors vs back-gate voltage **a** without light and with **b** 650 nm illuminations. **c** Light power-depended responsivity and shift of Dirac point under 650 nm illumination. **d** Light power-depended responsivities at different illumination wavelengths. Reproduced with permission [18]. Copyright 2015, Springer Nature

potential barrier at the interface, resulting in a strong photogating effect and yielding a high responsivity of 78 A W^{-1} with fast response time of 80 μs. Zhang et al. [19] developed a broadband graphene/SWCNT Schottky junction photodetector with asymmetric electrodes by covering monolayer self-assembled SWCNTs with monolayer graphene. The as-fabricated device exhibited a good responsivity of 209 mAW^{-1} and detectivity of 4.87×10^{10} Jones with short rise/fall time of 68/78 μs in a wide range of switching frequencies (50–5400 Hz). On the other hand, C_{60} was introduced into the graphene/SWCNT vdW heterostructures to form all-carbon nanohybrids with double vdW interface, in which the C_{60} and SWCNT are used as strong absorption in the visible and NIR region [48]. The double all-carbon vdW interface photodetector has been demonstrated to exhibit high responsivity up to >10^4 A W^{-1} and 545.7 A W^{-1} under 450 and 780 nm illuminations, respectively. Interestingly, the response time of the device toward NIR lights could be modulated to reduce an order of magnitude by visible light as gate signal.

As we know, monolayer sc-SWCNT networks or ultrathin film with ultrahigh absorption coefficient still absorb a small fraction of the incident lights, resulting in a low responsivity. While the increase of the responsivity is limited by increasing the SWCNT thickness or density due to the short exciton diffusion length perpendicular to the films or networks [49, 50]. To obtain high light detection ability (or detectivity), dark current of the device has to be further suppressed, which makes continuous graphene need to be redesigned in all-carbon vdW heterostructure photodetectors. Cai et al. [46] developed a novel all-carbon vdW heterostructure photodetector by spin-coating sc-SWCNT networks on the surface of separated graphene nanosheets, as shown in Fig. 6.5. The dark current can be controlled by changing the density of sc-SWCNT networks, thereby improving the detectivity of the all-carbon vdW heterostructure photodetectors. Moreover, the channel number of sc-SWCNTs connected to the electrodes and the density of all-carbon vdW heterostructures in the conducting channel can be controlled by non-uniform arrangement techniques, giving a significant enhancement for both responsivity and detectivity. Under light illuminations (405–1064 nm), the all-carbon vdW heterostructure photodetector exhibited a broadband photoresponse with a high responsivity of >3 × 10^3 A W^{-1} and a fast response speed of 44 μs, still representing the best result for SWCNT-based photodetectors so far (Table 6.2). Meanwhile, the detectivity up to 4.25 × 10^{12} Jones is comparable to those of pervious SWCNT-based photodetectors [18, 19, 47, 51].

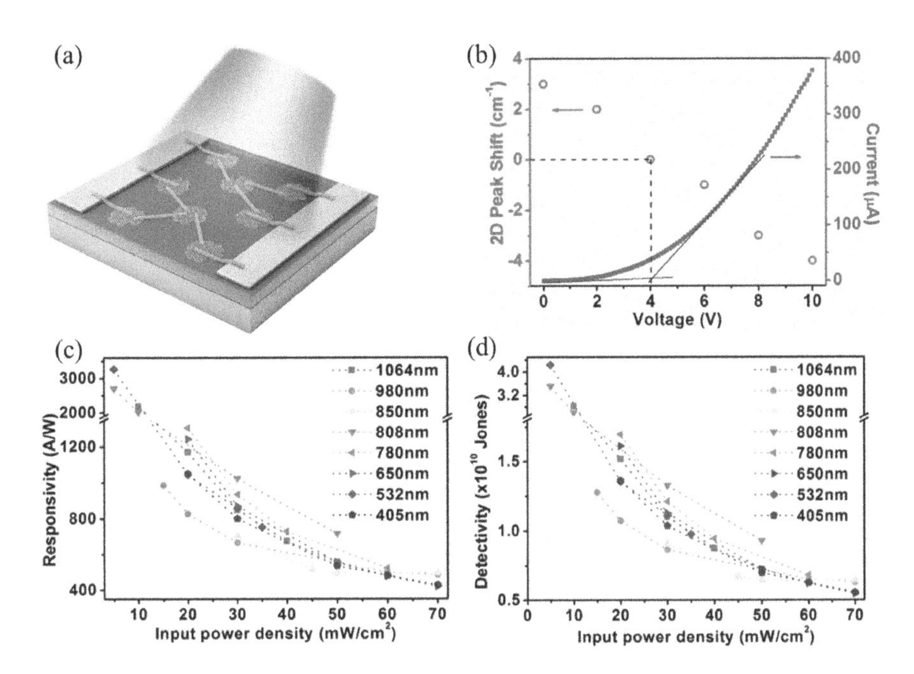

Fig. 6.5 **a** Schematic diagram of a photodetector based on sc-SWCNT/separated graphene vdW heterostructures. **b** I–V curve and 2D peak shifts of graphene in the heterostructures. **c** Responsivity and **d** detectivity versus the light power density. Reproduced with permission [46]. Copyright 2018, Wiley–VCH

Table 6.2 Comparison on the photoelectric performance of the SWCNT/graphene vdW heterojuntion photodetectors

Materials	Wavelength (nm)	Responsivity (A W^{-1})	Detectivity (Jones)	t_{rise}	t_{fall}	Ref
SWCNT/Gr	405 ~ 1550	>100	/	~ 100 μs	~ 100 μs	[18]
SWCNT/ GR	300 ~ 1100	0.209	4.87×10^{10}	68 μs	78 μs	[19]
SWCNT/ SGR	405 ~ 1064	>3000	$>4 \times 10^{10}$	44 μs	160 μs	[46]
SWCNT/ GR	Vis-NIR	78	/	80 μs	/	[47]
SWCNT/ GR/C_{60}	405–980	$>10^4$	/	~72 ms	/	[48]
SWCNT/ GR	405 ~ 650	51	/	40 ms	40 ms	[51]

In order to further improve the performance of SWCNT/graphene vdW junction photodetectors, novel device structures need to be designed, in which the high light absorption in sc-SWCNTs and low dark current can be obtained simultaneously. Figure 6.6 illustrates a conceptual device structure with all-carbon vdW junctions. The sc-SWCNT networks or films as photo active materials are firstly deposited and patterned on the substrate, then monolayer or few-layer graphene is transferred on the surface to the patterned sc-SWCNT networks or films. The dark current of the all-carbon vdW junction photodetectors can be controlled by changing the channel number of sc-SWCNT or m-SWCNTs aligned by high-frequency AC dielectrophoresis. The light absorption ability of sc-SWCNTs networks or films can be enhanced by increasing the network density or the film thickness of sc-SWCNTs.

6.7 Summary and Outlook

In summary, the formation of all-carbon heterostructures has been introduced firstly, and then we summarize systematically the state-of-the-art research progress of all-carbon vdW heterojunction photodetectors, in which C_{60}/SWCNT, C_{60}/graphene, carbon QD/graphene and SWCNT/graphene all-carbon heterojunctions are used active materials, respectively. Obviously, the all-carbon vdW heterojunctions have been demonstrated to exhibit exciting photoelectric performance, some of key parameters are several orders of magnitude higher than those of traditional commercial photodetectors, implying a great potential for future applications. However, it is worth noting that many challenges have to be overcome before the true commercial applications. Much more investigations are considered to be focused in the following aspects.

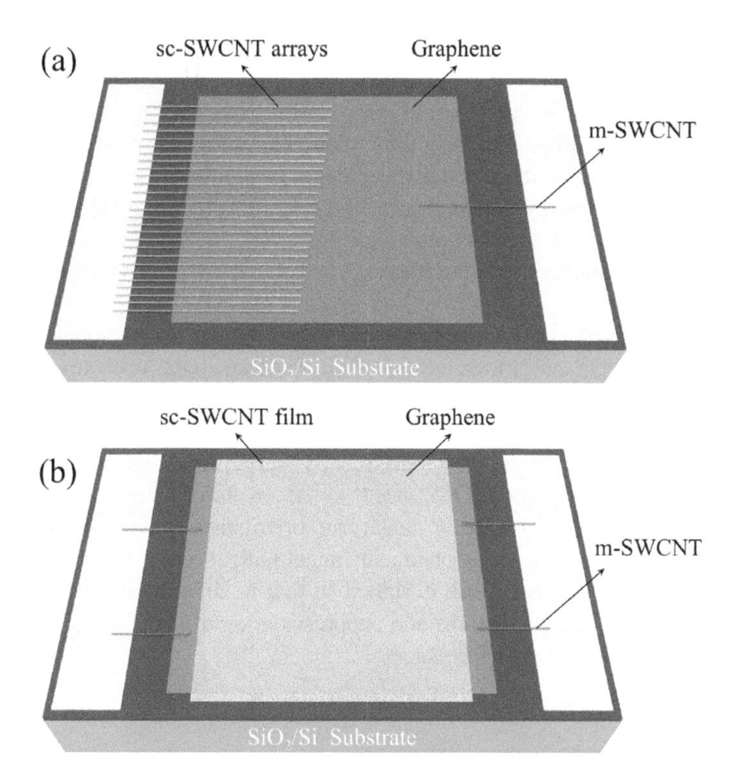

Fig. 6.6 Schematic diagram of all-carbon vdW junction photodetectors based on **a** orientated sc-SWCNT array/graphene and **b** sc-SWCNT film/ graphene, in which the dark current in controlled by changing the number of m-SWCNT channels

Firstly, the basic physical mechanism of all-carbon vdW heterojunction needs to be further revealed deeply. As we know, the contact type and area in the C_{60}/SWCNT, C_{60}/graphene and SWCNT/graphene heterostructures are very different from each other, and all of them are different from those in the typical 2D/2D vdW heterostructures. Thus, the heterojunction-induced electron coupling effect and interface carrier dynamic process are not still investigated systemically, which will play a critical role in designing all-carbon vdW heterojunction photodetectors. On the other hand, both SWCNTs and graphene possess unique electron band structures, which make the enhancement of the built-in field at the interface more flexible by regulating the Fermi level of sc-SWCNTs and graphene. In addition, the light-heterojunction interaction and the exciton dissociation in sc-SWCNTs also need to be paid much more attentions.

Secondly, high-quality carbon nanomaterials are still one of main obstacles for high performance photodetectors. For sc-SWCNTs, high purified single-chirality sc-SWCNTs are always needed during SWCNT-based nanodevices, especially for photodetectors. The presence of sc-SWCNTs with other chirality or m-SWCNTs will decrease the efficiency of free carrier collection due to the existence of trapped excitons or recombination in the intertube junctions, thereby the sensitivity and speed of photodetectors will be reduced significantly. For graphene, large-area defect-free graphene is very important for graphene-based all-carbon vdW heterojunction photodetectors. Especially, many broken pieces or structural defects are easily formed during the wet transfer process of graphene, severely limiting the efficient charge transport of photogenerated carriers through graphene. Meanwhile, the clean surface of graphene is also highly desired for the wet transfer process, which directly influences the interface carrier dynamic process and the device performance.

Thirdly, although simple all-carbon vdW heterojunction photodetectors with excellent photoelectric performance for point detection have been investigated in the past few years, there is still a long way toward practical application. The most important thing is to develop the photodiode array detectors using abovementioned all-carbon vdW heterojunctions. Consequently, large area, uniform thin film of sc-SWCNTs, C_{60} or carbon QDs need to be explored firstly, especially for sc-SWCNT film with controllable density or orientation. Therefore, much basic works need to be done, such as chemical functionalization or alignment. In addition, the device structures are still designed so that it can enhance light absorption of all-carbon vdW heterostructures and suppress dark current, thereby, resulting in an ultrahigh photoelectric performance.

References

1. Geim AK, Grigorieva IV (2013). Van der Waals heterostructures. Nature, 499(7459): 419–425.
2. Novoselov KS, Mishchenko A, Carvalho A, Neto AC (2016). 2D materials and van der Waals heterostructures. Science, 353(6298): aac9439.
3. Liu Y, Weiss NO, Duan X, Cheng HC, Huang Y, Duan X (2016). Van der Waals heterostructures and devices. Nature Rev Mater, 1(9): 1–17.
4. Jariwala D, Marks TJ, Hersam MC (2017). Mixed-dimensional van der Waals heterostructures. Nature Mater, 16(2): 170–181.
5. Su YJ, Xie MM, Lu XN, Wei H, Geng HJ, Yang Z, Zhang YF (2014). Facile synthesis and photoelectric properties of carbon dots with upconversion fluorescence using arc-synthesized carbon by-products. RSC Adv, 4(10): 4839–4842.
6. Dong XW, Su YJ, Geng HJ, Li ZL, Yang C, Li XL, Zhang YF (2014). Fast one-step synthesis of N-doped carbon dots by pyrolyzing ethanolamine. J Mater Chem C, 2(36): 7477–7481.
7. Shen J, Zhu Y, Yang X, Li C (2012). Graphene quantum dots: emergent nanolights for bioimaging, sensors, catalysis and photovoltaic devices. Chem Commun, 48(31): 3686–3699.
8. Yang L, Wang S, Zeng Q, Zhang Z, Peng LM (2013). Carbon nanotube photoelectronic and photovoltaic devices and their applications in infrared detection. Small, 9(8): 1225–1236.
9. He X, Léonard F, Kono J (2015). Uncooled carbon nanotube photodetectors. Adv Optical Mater, 3(8): 989–1011.
10. Qin SC, Liu YD, Jiang HZ, Xu YB, Shi Y, Zhang R, Wang FQ (2019). All-carbon hybrids for high-performance electronics, optoelectronics and energy storage. Sci China Inf Sci, 2019, 62: 220403.

11. Zhu Y, Li L, Zhang C, Casillas G, Sun Z, Yan Z, Ruan G, Peng Z, Raji AR, Kittrell C, Hauge RH (2012). A seamless three-dimensional carbon nanotube graphene hybrid material. Nature Commun, 3(1): 1–7.
12. Shi J, Li X, Cheng H, Liu Z, Zhao L, Yang T, Dai Z, Cheng Z, Shi E, Yang L, Zhang Z (2016). Graphene reinforced carbon nanotube networks for wearable strain sensors. Adv Funct Mater, 26(13): 2078–2084.
13. Kim SH, Song W, Jung MW, Kang MA, Kim K, Chang SJ, Lee SS, Lim J, Hwang J, Myung S, An KS (2014). Carbon nanotube and graphene hybrid thin film for transparent electrodes and field effect transistors. Adv Mater, 26: 4247–4252.
14. Shi E, Li H, Yang L, Hou J, Li Y, Li L, Cao A, Fang Y (2015). Carbon nanotube network embroidered graphene films for monolithic all-carbon electronics. Adv Mater, 27(4): 682–688.
15. Wang R, Hong T, Xu YQ (2015). Ultrathin single-walled carbon nanotube network framed graphene hybrids. ACS Appl Mater Interfaces, 7(9): 5233–5238.
16. Yan Z, Peng Z, Casillas G, Lin J, Xiang C, Zhou H, Yang Y, Ruan G, Raji ARO, Samuel EL, Hauge RH, Yacaman MJ, Tour JM (2014). Rebar graphene. ACS Nano, 8(5): 5061–5068.
17. Lv R, Cruz-Silva E, Terrones M (2014). Building complex hybrid carbon architectures by covalent interconnections: graphene–nanotube hybrids and more. ACS Nano, 8(5): 4061–4069.
18. Liu Y, Wang F, Wang X, Wang X, Flahaut E, Liu X, Li Y, Wang X, Xu Y, Shi Y, Zhang R (2015). Planar carbon nanotube–graphene hybrid films for high-performance broadband photodetectors. Nature Commun, 6: 8589.
19. Zhang TF, Li ZP, Wang JZ, Kong WY, Wu GA, Zheng YZ, Zhao YW, Yao EX, Zhuang NX, Luo LB (2016). Broadband photodetector based on carbon nanotube thin film/single layer graphene Schottky junction. Sci Rep, 6: 38569.
20. Yang Y, Yang X, Liang L, Gao Y, Cheng H, Li X, Zou M, Ma R, Yuan Q, Duan XF (2019). Large-area graphene-nanomesh/carbon-nanotube hybrid membranes for ionic and molecular nanofiltration. Science, 364(6445): 1057–1062.
21. Chae SH, Yu WJ, Bae JJ, Duong DL, Perello D, Jeong HY, Ta QH, Ly TH, Vu QA, Yun M, Duan XF, Lee YH (2013). Transferred wrinkled Al_2O_3 for highly stretchable and transparent graphene–carbon nanotube transistors. Nature Mater, 12(5): 403–409.
22. Seo H, Yun HD, Kwon SY, Bang IC (2016). Hybrid graphene and single-walled carbon nanotube films for enhanced phase-change heat transfer. Nano Lett, 16(2): 932–938.
23. Dowgiallo AM, Mistry KS, Johnson JC, Blackburn JL (2014) Ultrafast spectroscopic signature of charge transfer between single-walled carbon nanotubes and C_{60}. ACS Nano 8 (8): 8573–8581.
24. Pfohl M, Glaser K, Graf A, Mertens A, Tune DD, Puerckhauer T, Alam A, Wei L, Chen Y, Zaumseil J, Colsmann A (2016). Probing the diameter limit of single walled carbon nanotubes in SWCNT: fullerene solar cells. Adv Energy Mater, 6(21):1600890.
25. Classen A, Einsiedler L, Heumueller T, Graf A, Brohmann M, Berger F, Kahmann S, Richter M, Matt GJ, Forberich K, Zaumseil J (2019). Absence of charge transfer state enables very low V_{OC} losses in SWCNT: fullerene solar cells. Adv Energy Mater, 9(1): 1801913.
26. Bergemann K, Léonard F (2018). Room-temperature phototransistor with negative photoresponsivity of 10^8 AW^{-1} using fullerene-sensitized aligned carbon nanotubes. Small, 14: 1802806.
27. Park S, Kim SJ, Nam JH, Pitner G, Lee TH, Ayzner AL, Wang HL, Fong SW, Vosgueritchian M, Park YJ, Brongersma ML, Bao ZN (2015). Significant enhancement of infrared photodetector sensitivity using a semiconducting single-walled carbon nanotube/C_{60} phototransistor. Adv Mater, 27: 759–765.
28. Qin S, Chen X, Du Q, Nie Z, Wang X, Lu H, Wang X, Liu K, Xu Y, Shi Y, Zhang R, Wang F (2018). Sensitive and robust ultraviolet photodetector array based on self-assembled graphene/C_{60} hybrid films. ACS Appl Mater Interfaces, 10(44): 38326–38333.

29. Chugh S, Adhikari N, Lee JH, Berman D, Echegoyen L, Kaul AB (2019). Dramatic enhancement of optoelectronic properties of electrophoretically deposited C_{60}-graphene hybrids. ACS Appl Mater Interfaces, 11(27): 24349–24359.
30. Qin S, Jiang H, Du Q, Nie Z, Wang X, Wang W, Wang X, Xu Y, Shi Y, Zhang R, Wang F (2019). Planar graphene-C_{60}-graphene heterostructure for sensitive UV-visible photodetection. Carbon, 146: 486–490.
31. Pan R, Han JY, Zhang XC, Han Q, Zhou HX, Liu XC, Gou J, Jiang YD, Wang J (2020). Excellent performance in vertical graphene-C_{60}-graphene heterojunction phototransistors with a tunable bi-directionality. Carbon, 162: 375–381.
32. Yu X, Dong Z, Yang JKW, Wang QJ (2016). Room-temperature mid-infrared photodetector in all-carbon graphene nanoribbon-C_{60} hybrid nanostructure. Optica, 3(9): 979–984.
33. Yan X, Li B, Li LS (2013). Colloidal graphene quantum dots with well-defined structures. Acc Chem Res, 46(10): 2254–2262.
34. Yan Y, Gong J, Chen J, Zeng Z, Huang W, Pu K, Liu J, Chen P (2019). Recent advances on graphene quantum dots: from chemistry and physics to applications. Adv Mater, 31(21): 1808283.
35. Lim SY, Shen W, Gao Z (2015). Carbon quantum dots and their applications. Chem Soc Rev, 44(1): 362–381.
36. Cheng SH, Weng TM, Lu ML, Tan WC, Chen JY, Chen YF (2013). All Carbon-based photodetectors: An eminent integration of graphite quantum dots and two dimensional graphene. Sci Rep, 3: 2694.
37. Tetsuka H (2017). 2D/0D graphene hybrids for visible-blind flexible UV photodetectors. Sci Rep, 7: 5544.
38. Tetsuka H, Nagoya A, Tamura SI (2016). Graphene/nitrogen-functionalized graphene quantum dot hybrid broadband photodetectors with a buffer layer of boron nitride nanosheets. Nanoscale, 8: 19677–19683
39. Fantuzzi P, Candini A, Chen Q, Yao XL, Dumslaff T, Mishra N, Coletti C, Müllen K, Narita A, Affronte M (2019). Color sensitive response of graphene/graphene quantum dot phototransistors. J Phys Chem C, 123(43): 26490–26497.
40. Kim CO, Hwang SW, Kim S, Shin DH, Kang SS, Kim JM, Jang CW, Kim JH, Lee KW, Choi SH, Hwang E (2015). High-performance graphene-quantum-dot photodetectors. Sci Rep, 4: 5603.
41. Haider G. Roy P. Chiang CW. Tan WC. Liou YR. Chang HT. Liang CT. Shih WH. Chen YF (2016). Electrical-polarization-induced ultrahigh responsivity photodetectors based on graphene and graphene quantum dots. Adv Funct Mater, 26(4): 620–628.
42. Lu R, Christianson C, Weintrub B, Wu JZ (2013). High photoresponse in hybrid graphene-carbon nanotube infrared detectors. ACS Appl Mater Interfaces, 5(22): 11703–11707.
43. Rao R, Pierce N, Dasgupta A (2014). On the charge transfer between single-walled carbon nanotubes and graphene. Appl Phys Lett, 105: 073115.
44. Liu Y, Wang F, Liu Y, Wang X, Xu Y, Zhang R (2016). Charge transfer at carbon nanotube-graphene van der Waals heterojunctions. Nanoscale, 8: 12883–12886.
45. Cai BF, Yin H, Huo TT, Ma J, Di ZF, Li M, Hu NT, Yang Z, Zhang YF, Su YJ (2020). Semiconducting single-walled carbon nanotube/graphene van der Waals junctions for highly sensitive all-carbon hybrid humidity sensors. J. Mater Chem C, 8: 3386–3394.
46. Cai BF, Su YJ, Tao ZJ, Hu J, Zou C, Yang Z, Zhang YF (2018). Highly sensitive broadband single-walled carbon nanotube photodetectors enhanced by separated graphene nanosheets. Adv Optical Mater, 6(23): 1800791.
47. Cao J, Zou Y, Gong X, Gou P, Qian J, Qian R, An Z (2018). Double-layer heterostructure of graphene/carbon nanotube films for highly efficient broadband photodetector. Appl Phys Lett, 113(6): 061112.
48. Zhou HX, Yang M, Ji CH, Liu XC, Pan R, Han Q, Wang J, Jiang YD (2020). Excellent-performance C_{60}/graphene/SWCNT heterojunctionwith light-controlled enhancement of photocurrent. ACS Sustainable Chem Eng, 8(10): 4276–4283.

49. GI Koleilat, M Vosgueritchian, T Lei, Y Zhou, DW Lin, F Lissel, P Lin, JW To, T Xie, K England, Y Zhang, Bao ZN (2016). Surpassing the exciton diffusion limit in single-walled carbon nanotube sensitized solar cells. ACS Nano, 10(12): 11258–11265.
50. Dowgiallo AM, Mistry KS, Johnson JC, Reid OG, Blackburn JL (2016). Probing exciton diffusion and dissociation in single-walled carbon nanotube-C_{60} heterojunctions. J Phys Chem Lett, 7(10): 1794–1799.
51. Liu Y, Liu Y, Qin S, Xu Y, Zhang R, Wang F (2017). Graphene-carbon nanotube hybrid films for high-performance flexible photodetectors. Nano Res, 10, 1880–1887.

Chapter 7
Carbon Nanotube/semiconductor van der Waals Heterojunction Solar Cells

7.1 Introduction

Solar cells convert sunlight into electricity based on photovoltaic effects, giving a clean and renewable energy in our daily life. Wafer-based bulk semiconductor (Si, GaAs, etc.) solar cells always dominate the solar cell commercial market due to mature material and manufacturing process as well as superior stability. At present, the records of power conversion efficiency (PCE) for single-junction and multi-junction solar cells reach 29.1 and 39.2% under the global AM 1.5 spectrum [1]. Although the PCE of solar cells is still increasing by a small margin through optimizing the material process and absorbing sunlight by more sub-solar cells, a significant improvement of the PCE is still a big challenge. Therefore, highly efficient harvest of solar energy has drawn continuous attention with the development of nanomaterials and nanotechnology over the last few decades. Especially, carbon nanotube (CNT) as a typical quasi-1D nanomaterial possesses outstanding optical, electrical properties and excellent chemical stability due to its unique carbon-atom sp^{n} hybridization [2, 3]. The electron ballistic transport along the axial direction of CNTs enables high carrier transportation capability, and the electron mobility is as high as 10^5 cm^2 V^{-1} s^{-1} at room temperature. Importantly, single-walled CNTs (SWCNTs) exhibit highly chirality-depended optical and electrical properties. One third of SWCNTs behave metallic, and the remaining two-thirds are weak p-type direct-bandgap semiconductors in air with a bandgap inversely proportional to their diameter. Moreover, high absorption coefficient (10^4–10^5 cm^{-1}) with UV-NIR light absorption enable semiconducting SWCNTs (sc-SWCNTs) to be an ideal light-harvesting material for next-generation high efficient photovoltaic cells [4–6].

Among the photoelectric devices based on CNTs and CNT-based heterojunctions, the CNT/semiconductor vdW heterojunction solar cells have attracted considerable interest due to their simple structure, easy fabrication and low cost. In these devices, the metallic CNT films as transparent conductive electrodes are

Y. Su, *High-Performance Carbon-Based Optoelectronic Nanodevices*,
Springer Series in Materials Science 319,
https://doi.org/10.1007/978-981-16-5497-8_7

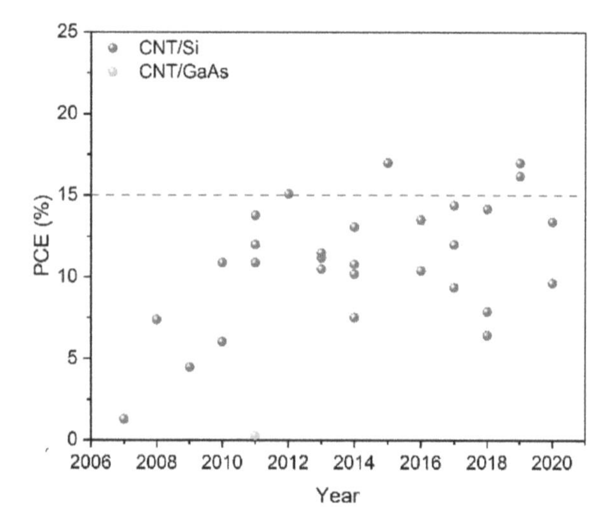

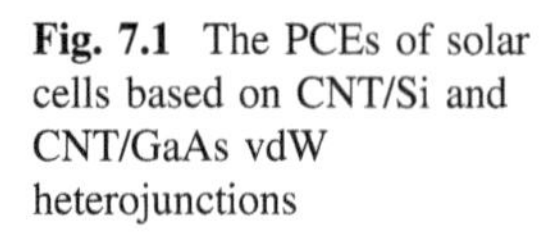
Fig. 7.1 The PCEs of solar cells based on CNT/Si and CNT/GaAs vdW heterojunctions

mainly used to separate and transport the photogenerated carriers at the CNT/semiconductor interface due to the formation of a built-in potential. However, sc-SWCNTs do not separate and transport the photogenerated carriers, but also contribute a certain photocurrent through light absorption. In the past ten years, many efforts have been made to improve the PCE of CNT/semiconductor vdW heterojunction solar cells by optimizing the device structure, interface passivation, and chemical doping, and so on [7–12]. The PCE of CNT/semiconductor solar cells has increased rapidly to over 15% in the past few years (Fig. 7.1). In this chapter, we focus on the CNT/semiconductor vdW heterojunctions solar cells. Firstly, the work mechanism and key parameters will be introduced, and then the recent research progress on CNT/Si and CNT/GaAs solar cells will be systemically summarized, where SWCNT, DWCNT and MWCNT are used. Finally, the current challenges and future perspectives of CNT/semiconductor vdW heterojunction solar cells will be discussed.

7.2 Work Mechanism and Key Parameters

7.2.1 PN and Schottky Heterojunctions

As we know, SWCNTs can behave semiconducting or metallic according to their chiralities and diameters, the p-n vdW heterojunctions will be formed when sc-SWCNTs contact with bulk semiconductors and a Schottky heterojunction will be formed by replacing sc-SWCNTs with m-SWCNTs. However, both type vdW heterojunctions usually play a role in determining the working mechanism for the

CNT/semiconductors consisted of mixed SWCNTs or MWCNTs. Thus, taking n-type silicon as an example, two type vdW heterojunctions will be respectively introduced in the following.

When sc-SWCNTs contact with n-type Si, an inter-diffusion process of charge carriers occurs due to the concentration difference of electrons and holes, in which the electrons (holes) diffuse from Si (or sc-SWCNTs) to sc-SWCNTs (or Si). Due to the formation of a built-in electric field at the interface, the energy band of sc-SWCNTs and n-Si bends until the thermal equilibrium (Fig. 7.2a), and similar band bending also happens in Schottky junction assembled by metallic SWCNTs and n-type Si (Fig. 7.2b). Under light illuminations, the thermal equilibrium is broken in the heterojunctions (Fig. 7.2c). For SWCNT/Si p-n junctions, the incident photons are absorbed simultaneously by sc-SWCNTs and Si, a large number of electron-hole pairs are excited by the photons with energy larger than the bandgap. And then, the photogenerated electron–hole pairs are separated at the interface under the built-in field. The corresponding open voltage of solar cells is equal to the difference between the Fermi levels of SWCNT and Si. Obviously, a larger built-in field of the p-n heterojunctions is benefits to improve the collection of photogenerated carriers by tailoring the work function of sc-SWCNTs. Likewise,

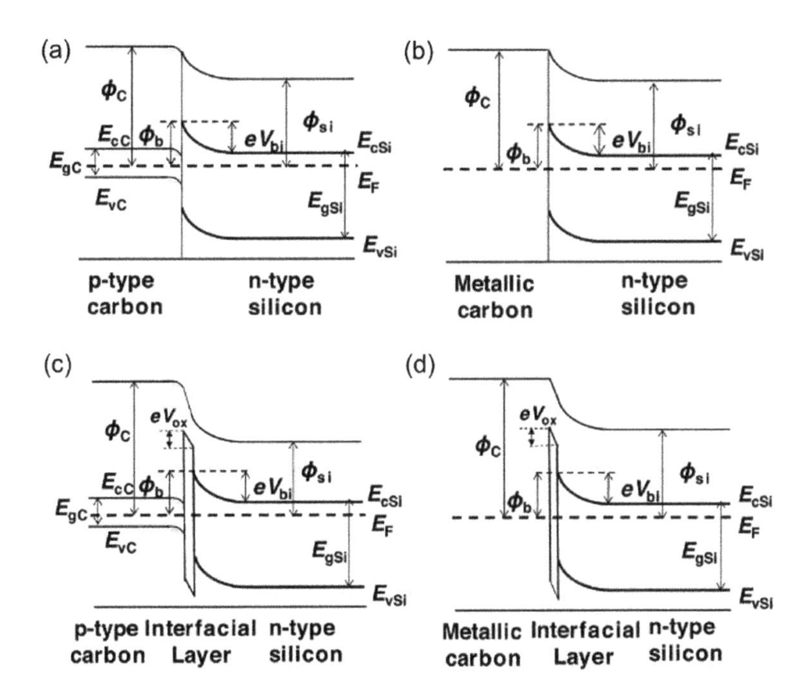

Fig. 7.2 Energy band diagrams of the SWCNT/Si heterojunctions. **a**, **c** p/n heterojunctions and **b**, **d** Schottky heterojunctions without and with interfacial modification. Reproduced with permission [4]. Copyright 2015, Wiley

the built-in field of the Schottky heterojunctions can also be enhanced by controlling the work function of metallic SWCNTs (Fig. 7.2d), resulting in a high collection efficiency of photogenerated carriers.

7.2.2 Key Parameters of Solar Cells

As a photovoltaic device, the main characteristic parameters are power conversion efficiency (PCE, η), open voltage (V_{OC}), short-circuit current density (J_{SC}), Fill factor (FF), external quantum efficiency (EQE) and internal quantum efficiency (IQE). According to the definition of PCE, it is usually expressed as:

$$\eta = \frac{J_{sc} \times V_{oc} \times FF}{P_{in}} \tag{7.1}$$

where P_{in} is the incident light power density (100 mW/cm^2 under AM 1.5 G condition). Clearly, in order to promote the PCE in the CNT/semiconductor solar cells, the main point is to improve at least one of three parameters by controlling the Fermi level of CNTs, interface contact or surface passivation. For an ideal solar cell, the equivalent circuit can be modeled as a current source in anti-parallel with a diode. However, the effects of series (R_s) and shunt (R_{sh}) resistance have to be considered in the realistic solar cells, in which the R_s corresponds to the bulk and contact resistances and the shunt resistance consists of all parallel resistive losses across the cell. Correspondingly, the voltage-current characteristics of solar cells can be expressed as:

$$I = I_{ph} - I_0 e^{\frac{e(V+IR_s)}{nkT}} - \frac{V+IR_s}{R_p} \tag{7.2}$$

where n is the ideal factor, and I_0 is the reverse saturation current of the junction. R_s and R_p have distinct impact on the FF. Obviously, the change of R_s does not affect the Voc value, while the decrease of R_p significantly reduces the Voc value and thereby results in a low PCE.

Expect for the parameters from the J-V curves, the quantum efficiency (EQE and IQE) is also very important to evaluate the photoelectric performance of solar cells, which is mainly used to characterize the number of photogenerated electron-hole pairs per incident/absorbed photons with a given energy. The IQE is determined by the absorbed photons which need to deduce the photon losses from the total incident photons due to the reflection (R) and transmission (T). Thus, the relationship between EQE and IQE can be expressed as:

$$\mathrm{IQE} = \mathrm{EQE}/(1 - \mathrm{R} - \mathrm{T}) \approx \mathrm{EQE}/(1 - \mathrm{R}) \tag{7.3}$$

From the EQE and IQE curves (Fig. 7.4), one can better understand the light loss mechanism of solar cells, which can be used to evaluate the front surface combination, the absorption efficiency of different photons, and non-absorbable spectral region. To improve the efficiency of the CNT/semiconductor vdW heterojunction solar cells, more attention has also be paid to the full utilization of high- and low-energy photons and the design of ultra-low reflection layer.

7.3 DWCNT or MWCNT/Si vdW Heterojunction Solar Cells

The CNT/Si heterojunction solar cells have been firstly proposed by coating semi-transparent DWCNT film on n-type silicon to create p-n heterojunctions in 2007, in which DWCNTs served as both photogeneration sites and a collecting/transport layer of charge carriers [7]. Importantly, the distinct properties of silicon (e.g., diffusion length) and CNTs (e.g., high mobility) could be fully utilized to construct highly efficient solar cells. Although the PCE of device is only about 1%, the vdW heterojunction solar cells demonstrated a potentially suitable configuration for novel solar cells through a simple and scalable process. Jia et al. [13] further optimized the DWCNT/Si solar cells by considering bandgap, Fermi level and thickness of spiderweb-like DWCNT films, The solar cell showed a Jsc of 26 mA/cm^2, Voc of 0.54 V, a FF of 53%, and a PCE of 7.4%. Moreover, the separation of metallic and semiconducting nanotubes was not required during the fabrication process. By introducing a thin oxide layer between MWCNT and Si, the PCE of MWCNT/insulator/Si heterojunction solar cell could be further increased to 10.1% owing to reduced internal resistance and suppressed charge recombination [14]. In the past ten years, many efforts have been made in the efficiency improvements of CNT/Si vdW solar cells, which can be divided into two main aspects: the conductivity enhancement and Fermi level control of CNT films.

7.3.1 Improving the Conductivity of the CNT Films

Since the electrical conductivity and transparency of CNT thin films play an important role in the carrier separation, transfer and collection and solar cell performance, the optimization of CNT films has always attracted attentions during the fabrication of CNT/Si solar cells. Di et al. [15] demonstrated that well-aligned DWCNT films could enhance the conversion efficiency of CNT/Si Schottky solar cells, in which uniform and dense Schottky heterojunctions were more easily formed and photogenerated charges could be transported efficiently along one direction on the DWCNTs. Compared with the pristine and random DWCNT networks, the device had been achieved higher PCE of 10.5% with a Jsc of

33.4 mA/cm^2, V_{OC} of 0.54 V and FF of 0.58. Gan et al. [16] synthesized the MWCNT-graphene hybrid sheets with high transparency, excellent electrical conductivity and mechanical strength. Using the hybrid film with chemical doping, the power conversion efficiency up to 8.50% has been demonstrated in the graphene-MWCNT/Si solar cells. Xu et al. [11] firstly developed a large-area MWCNT/Si solar cell (>2 cm^2) by improving the carrier collection efficiency of CNT film with flattened highly conductive MWCNT strips (Fig. 7.3), in which MWCNT strips play a role like traditional metal grids without introducing contact barrier. The power conversion efficiency of more than 10% has been realized by combining with antireflection and doping techniques.

Similar approach has been used to improve the electric conductivity of the MWCNT films using poly(3, 4-ethylene dioxythiophene):poly(styrenesulfonate) (PEDOT:PSS) [17]. In the PEDOT:PSS-MWCNT hybrid film, PEDOT:PSS not only fills the nanometer scale pores of the MWCNT network, but also patches contact with Si concomitantly and through seamless contact with n-Si. The PCE of the heterojunction solar cell can be improved up to 10.2% (Fig. 7.4) due to the synergistic effects of the PEDOT:PSS-CNT hybrids.

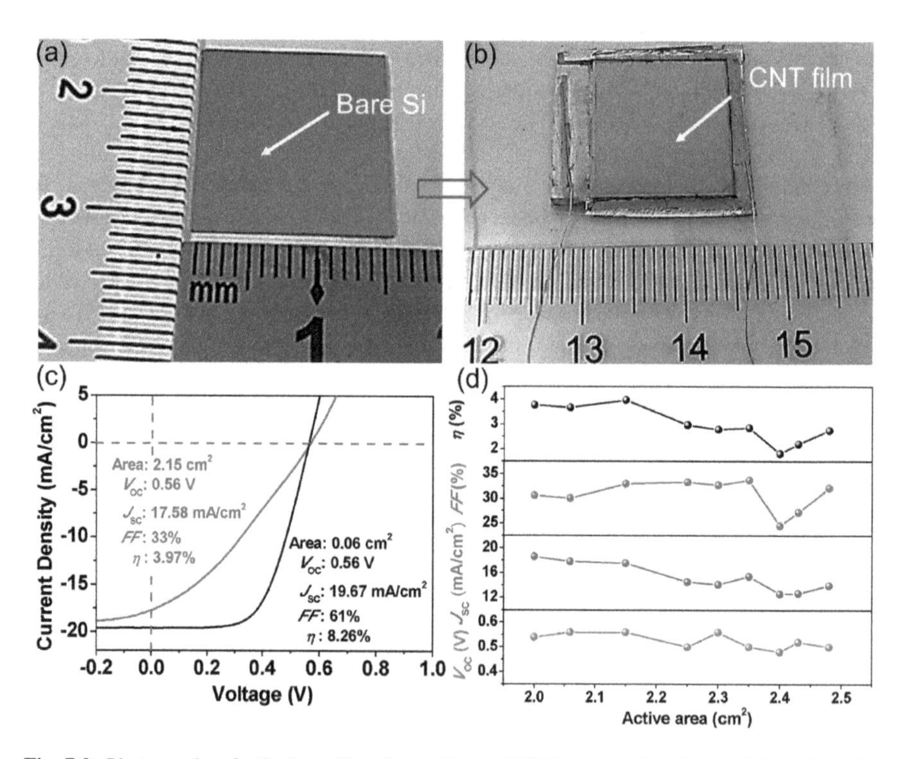

Fig. 7.3 Photographs of **a** the bare Si wafer and **b** the CNT/Si solar cell. **c** Comparision of the J-V characteristics of large and small area CNT/Si solar cells. **d** Parameters of large-area CNT/Si solar cells with different active areas. Reproduced with permission [11]. Copyright 2016, John Wiley and Sons

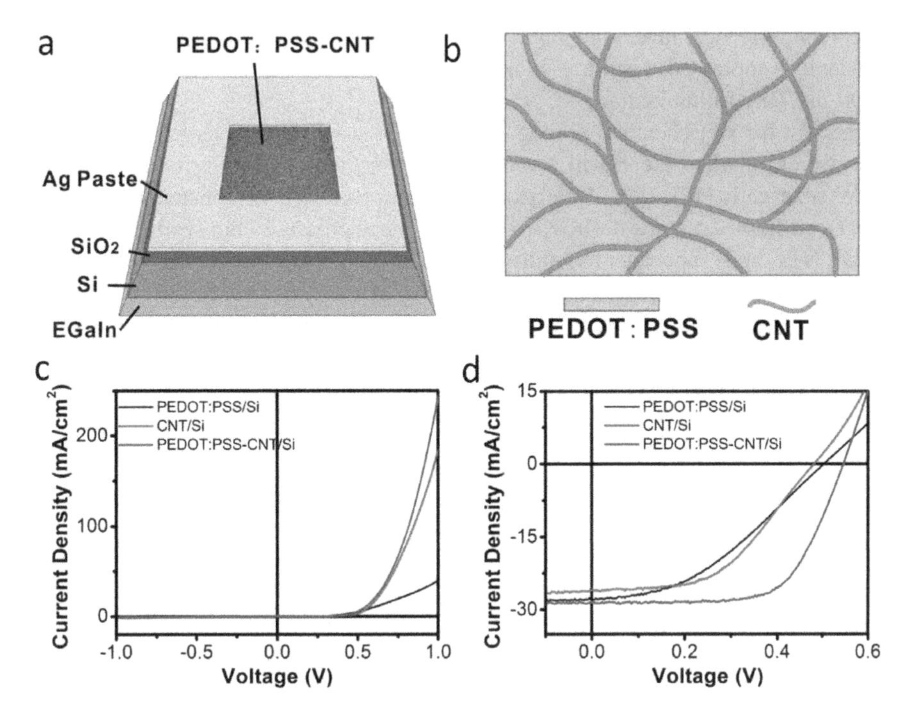

Fig. 7.4 **a** Schematic of the PEDOT:PSS-CNT/n-Si solar cell. **b** Diagram of the back surface of composite film. J-V characteristics of heterojunction solar cells **c** in the dark and **d** under AM 1.5 G illumination. Reproduced with permission [17]. Copyright 2017, Elsevier

7.3.2 Changing the Fermi Level of MWCNT Films

It has been demonstrated that chemical doping could control the Fermi level of MWCNTs which significantly influence the charge separation efficiency at the interface by changing the built-in field between MWCNTs and Si. For example, Wu et al. [18] demonstrated the significant improvement of the PCE of MWCNT/Si solar cells using certain metal chlorides, in which high V_{OC} values were usually obtained using the metal chlorides with large electrode potential of metal cations. These PCE enhancement can be attributed to the p-doping effect of MWCNTs and antireflection mechanisms of metal chlorides. Consequently, the MWCNT/Si solar cells with stable PCE of 16.2% have been achieved by spin-coating $ZrCl_4/FeCl_3$. Furthermore, transitional metal oxide (MoO_3 and WO_3) films transformed by their chlorides have also enables p-type doping on MWCNTs, resulting in a significant improvement in the photovoltaic performance of MWCNT/Si vdW solar cells. The efficiencies >16% could be also achieved by combining with other chlorides to increase Jsc [19]. In addition, Shi et al. [20] also reported a high efficiency CNT/Si vdW heterojunction solar cells using by HNO_3/H_2O_2 doped CNTs. By applying a

TiO_2 antireflection layer, the light reflectance from the Si surface has been significantly suppressed, resulting in an enhanced efficiency of 15.1% under AM 1.5 (100 mW/cm^2) illumination.

Except for wet chemical doping dry gas-doping strategy is very simple, highly effective to change the Fermi level CNTs during the fabricating of MWCNT/Si vdW solar cells [12]. Low-dose ozone treatment has been demonstrated to p-dope CNTs without structure destroy, resulting in an increase of the work function of MWCNTs and subsequently enhancement of the built-in electric field at the heterojunction. As a result, the PCE of the solar cell increased from 5.29% to 12.70% without degradation in 24 h due to the significant improvement of open-circuit voltage and fill factor, as shown in Fig. 7.5. Furthermore, high PCE of >17% can be achieved under AM 1.5 illumination conditions (100 mW/cm^2) by combining antireflection and acid doping.

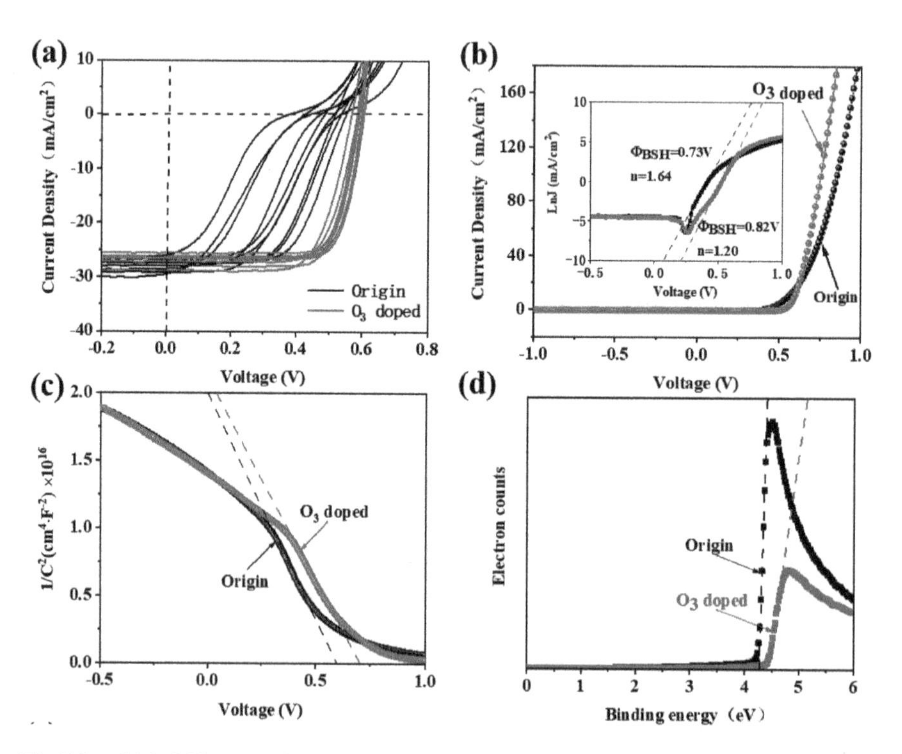

Fig. 7.5 **a** Light J-V curves of ten CNT/Si solar cells under AM 1.5 and **b** dark J-V curves before and after 5-min ozone exposure. **c** 1/C^2-V curves and **d** Ultraviolet photoelectron spectroscopy (UPS) valence-band structure of the CNT film before and after ozone exposure. Reproduced with permission [12]. Copyright 2019, Elsevier

7.4 SWCNT/Si vdW Heterojunction Solar Cells

Different from the MWCNTs with metallic properties, SWCNTs exhibit semiconducting or metallic properties according to its chirality and diameters. Especially, sc-SWCNT network coating on n-type silicon substrate exhibits a rectifying behavior owing to the formation of p-n heterojunctions. Under illumination, the electron-hole pairs are generated in both Si and sc-SWCNTs, which were then separated at the interface and transported through SWCNTs (holes) and n-Si (electrons), respectively. Therefore, sc-SWCNTs do not only serve as a carrier collecting and transport layer, but also contribute to the photocurrent [8]. Moreover, the barrier height between sc-SWCNTs and Si can be tuned by changing sc-SWCNTs with suitable chirality. In the past few years, various methods have been used to improve the efficiency of SWCNT/Si solar cells, [21–33]. After optimizing the heterojunction interface and the SWCNT films, the efficiency of SWCNT/Si solar cells has increased to over 15%.

7.4.1 Improving the Structure and Properties of SWCNT Films

In the SWCNT/Si vdW heterojunction solar cells, although the ultrathin p-n junctions have been formed using p-type SWCNTs and n-type Si, the thickness of SWCNTs is only few or tens of nanometers, resulting in a fact that the photocurrent contributed by SWCNTs is very low, even no contribution [34]. It has been demonstrated diffusion-dominated p-n junction transport and dominant carrier generation in Si for the SWCNT/Si p-n heterojunction solar cells [24]. Thereby, the main role of SWCNT film is to establish a built-in potential for separate the photogenerated electron/hole pairs and collect the holes generated in Si [24]. As we know, the sheet resistance of SWCNT film increases with the decrease of thickness at high transmittance, and the sheet resistance and transmittance of SWCNT film have a significant impact on the PCE of SWCNT/Si vdW solar cells. Therefore, it is very important to achieve a balance between sheet resistance and transmittance of SWCNT films.

The preparation techniques for SWCNT film are usually mainly classified into wet process [35–38] and dry process [39–41]. In the wet process, spin-coating can be used to directly deposit SWCNT network with a thin thickness on the substrate, while vacuum filtration is suitable to obtain a thick SWCNT film on the microporous filter and subsequently transfer to the target substrates. Because purification, dispersion, sonication and centrifugation are involved, the contamination and defects are inevitably introduced into the SWCNT films, resulting in a negative effect on the performance of SWCNT/Si solar cells. Li et al. [27] reported a wet-process preparation of smooth and aligned SWCNT films by utilizing a superacid slide casting method. Based on the SWCNTs with low surface roughness and high transparency, the solar cell exhibited showed a relatively high PCE of 8.52% with J_{sc} of 26.1 mA/cm^2, V_{oc} of 0.51 V, and FF of 0.64. The PCE of SWCNT/Si solar cells can be further optimized to be 11.2% with a high FF of 74.1% and an ideality factor close to unity (Fig. 7.6) [24]. Harris et al. [42] fabricated a SWCNT/Si vdW solar cell based on the freestanding (6, 5)-enriched sc-SWCNT films using a novel fluid-processing scheme, and a high PCE value of 14.85% was achieved in simple geometries without antireflective after chemical doping. Moreover, the results also suggested that these devices simultaneously exhibited characteristics of both Schottky and p-n junctions due to the present of impurity in the sc-SWCNTs.

Compared with the SWCNT film prepared by wet process, the dry-process SWCNT films usually have higher transparency and lower sheet resistance due to the easy control of film thickness, which is benefit to fabricate high performance SWCNT/Si solar cells. Cui et al. [9] demonstrated a SWCNT/Si solar cells by dry-transferring high crystallinity SWCNT film with long-bundle length (9.4 μm on average) on the surface of Si substrate, which was synthesized by floating catalyst CVD. Using the SWCNT film with a sheet resistance of 134 Ω/sq at 81.5% transmittance, the SWCNT/Si solar cell exhibited a high stability and a high PCE of

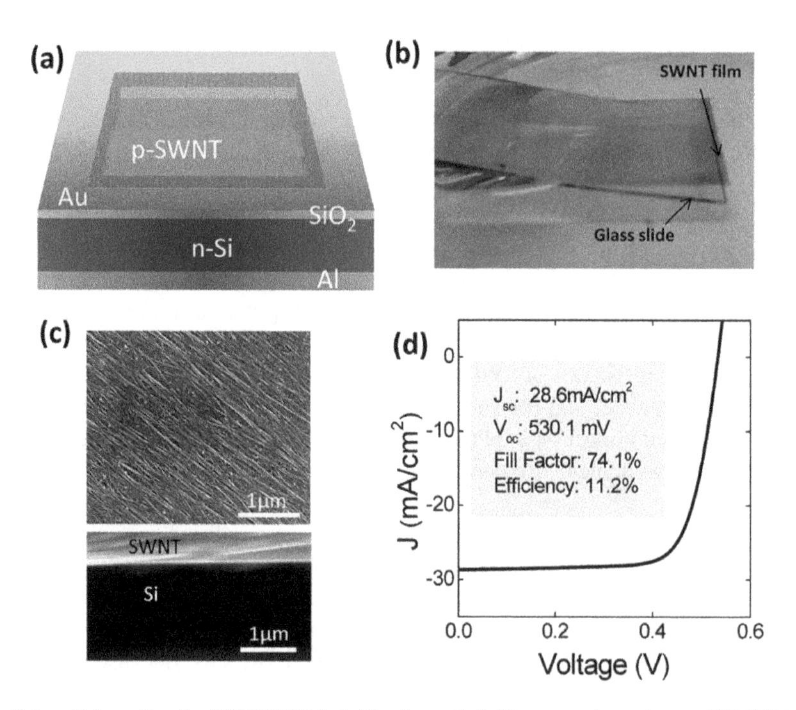

Fig. 7.6 **a** Schematic of a SWCNT/Si hybrid solar cell. **b** Photo-graph to show a SWCNT thin film prepared on a slide glass for transfer to show a Si wafer. **c** Plane-view of a representative SWCNT film and cross-sectioned view of SWCNT/Si interface. **d** J-V characteristics of a SWCNT/Si solar cell under 1 sun illumination. Reproduced with permission [24]. Copyright 2013, American Chemical Society

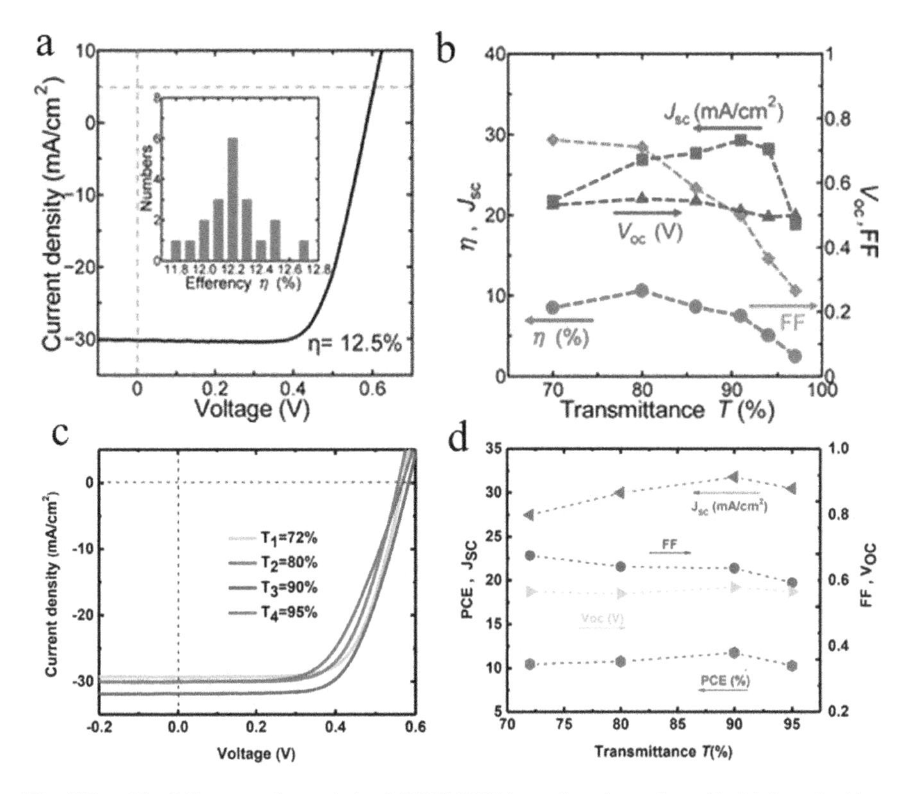

Fig. 7.7 **a** The J-V curves for optimized SWCNT/Si heterojunction solar cells fabricated with a window size of 1 mm using a SWCNT film with T = 91%. **b** Parameter values as a function of transmittance. Reproduced with permission [26]. Copyright 2014, American Chemical Society. **c** J–V curves of SWCNT/Si solar cells with SWCNT films of different thicknesses under illumination with a simulated AM 1.5 G solar light. **d** PCE, Voc, FF and Jsc values of solar cells as a function of transmittance. Reproduced with permission [31]. Copyright 2018, Elsevier

10.8% with a J_{sc} of 29.7 mA/cm^2, V_{oc} of 0.535 V, and FF of 0.68. Wang et al. [26] investigated the effect of film thickness on the PCE of SWCNT/Si solar cells, and found that the sheet resistance increased with an increase of transmittance of SWCNT films (Fig. 7.7a, b), resulting in that the current density J_{SC} increased first and then decreased with the increase of the transmittance of SWCNT film, while the FF always decreased. The PCE of SWCNT/Si solar cells with 1 mm diameter window reached the best value of 11.8–12.7% at 91% transmittance without post-processing. Hu et al. [31] demonstrated a silicon vdW heterostructure solar cell using a high-quality, small-bundle SWCNT film with a record low sheet resistance of 180 Ω/sq at 90% transmittance, which was synthesized by floating catalyst CVD rather than by self-assembly or filtration method. Consequently, the optimal PCE of solar cell with areas of ~9 mm^2 could be 11.8% at 90% transmittance and the PCE value decreased to 10.3% when the transmittance was further increased to 95% (Fig. 7.7c, d). At the optimal condition, the optimized SWCNT/Si vdW solar cells with areas of 2.3 mm^2 exhibited a high PCE value of 14.2% with a J_{sc} of 33.8 mA/cm^2, V_{oc} of 0.589 V and FF of 0.712 under AM 1.5 G illuminations. Cui et al. [25] proposed a novel SWCNT/Si heterojunction solar cell by self-assembling micro-honeycomb SWCNT network on the Si. Compared with random SWCNT films, this three-dimensional hierarchical structure exhibited lower sheet resistance and higher optical transmittance. Without optimizing the diameter or height of the vertically aligned SWCNTs, a PCE of 6% with a FF of 72% could be achieved, and the PCE could be improved to 10.02% with a FF of 73% after dilute nitric acid treatment.

In addition, to improve the conductivity of SWCNTs, Ag nanowires have been introduced on the surface of the SWCNT/Si vdW heterojunctions, [43] in which the SWCNT film with high density and low porosity were required to prevent directly contact between the Ag NWs and the Si. Although V_{oc} and J_{sc} decreased slightly, the FF increased significantly due to the decrease of the series resistance. A PCE of 10.8% with an active area of 49 mm^2 was achieved after depositing antireflective TiO_2 nanoparticles.

7.4.2 Optimizing the Fermi Level of SWCNT Films by Chemical Doping

The Fermi level of SWCNT plays an important role in improving the charge transfer between SWCNT and Si by changing the built-in electric field of the heterojunctions. The thionyl chloride ($SOCl_2$) modified SWCNTs have been demonstrated to enhance the charge separation at the SWCNT/Si interface through adjusting the Fermi level of p-type SWCNTs, [21] resulting in a significantly increase of the PCE of SWCNT/Si solar cells due to higher carrier concentration and mobility. Jia et al. [44] demonstrated that the PCE of SWCNT/Si heterojunction solar cells could be significantly improved to be 13.8% by HNO_3 doping, the results showed that the internal series resistance was reduced and more carrier transport paths in the porous SWCNT networks were generated after the HNO_3 doping. Hu et al. [32] demonstrated that lightly fluorinated treatment enabled improved electronic conductivity and a higher work function of SWCNT films due to the formation of ionic C–F bonds on tube walls with p-type doping effect. Meanwhile, the interface contacts between SWCNT and Si were improved due the areal density increased and the surface roughness decrease of SWCNT films during in the fluorination process. Consequently, the fluorinated SWCNT/Si solar cell exhibited a PCE of 13.6% and better long-term stability in air. Cui et al. [45] demonstrated an effective and scalable solid-state redox functionalization process for p-doping SWCNTs by coating colloidal $CuCl_2/Cu(OH)_2$ redox, as shown in Fig. 7.8. Owing to the electron transfer from SWCNT to the redox, the Fermi level of SWCNTs was moved towards the valence band. After $CuCl_2/Cu(OH)_2$ doping, the sheet resistance of the SWCNT films achieved 69.4 Ω/sq at 90% transparency, and the $CuCl_2$/Cu

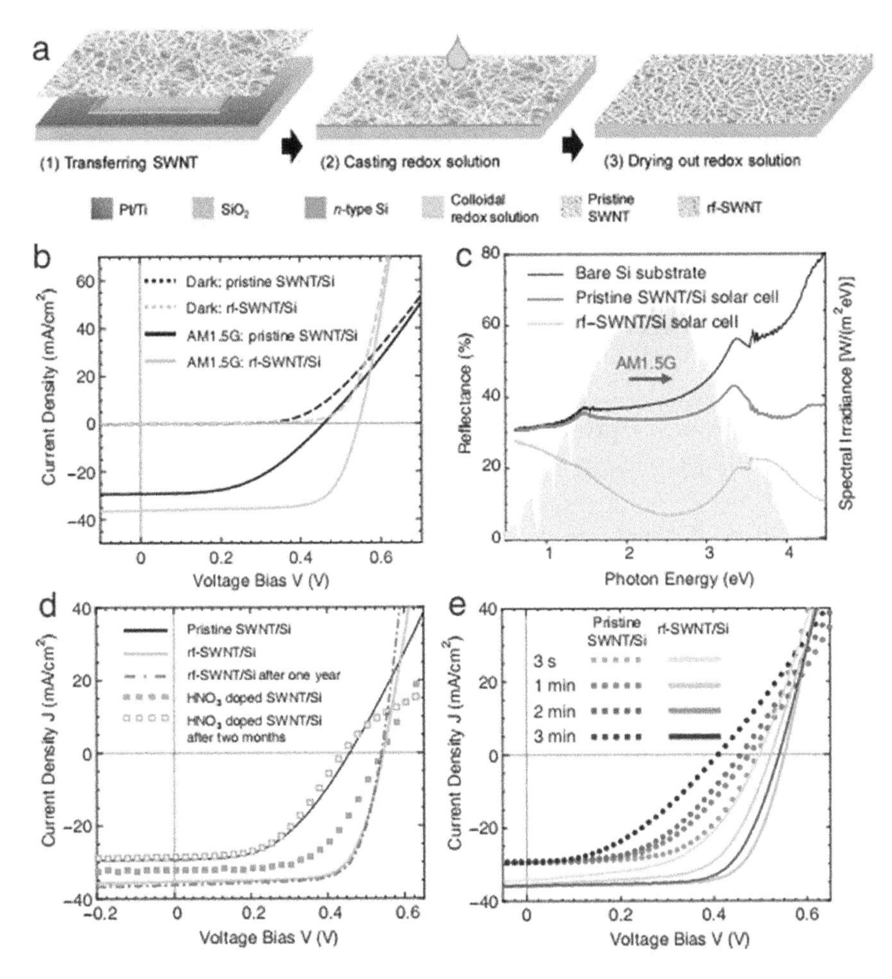

Fig. 7.8 **a** Schematics of the $CuCl_2/Cu(OH)_2$ redox functionalization process for the SWCNT/Si solar cells. **b** J-V characteristics of the SWCNT/Si solar cells before and after the $CuCl_2/Cu(OH)_2$ redox functionalization process. **c** UV-vis-NIR reflectance spectra of bare Si, the pristine SWCNT film on bare Si and the rf-SWCNT film on bare Si. **d** Comparison of the J-V characteristics of the pristine SWCNT/Si solar cell. **e** Effect of varying RCA treatment time on the J-V characteristics of the pristine SWCNT/Si solar cells and the rf-SWNT/Si solar cells. Reproduced with permission [45]. Copyright 2017, Wiley

$(OH)_2$ redox coating also served as an antireflection layer, which is helpful for the photocurrent increase in the SWCNT/Si vdW heterojunction solar cell. As a result, the PCE of the device after redox functionalization increases significantly from 6.59 to 13.77%. Importantly, the value reached 14.09% with more than one-year stability in ambient environment.

Qian et al. [46] demonstrated an enhancement of the photovoltaic performance of SWCNT/Si solar cells by doping SWCNT film using a polymeric acid (Nafion),

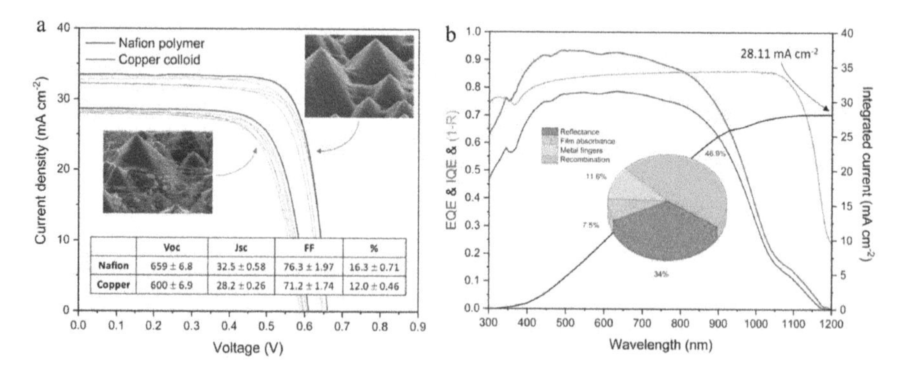

Fig. 7.9 **a** Performance of solar cells doped with either copper colloid or polymeric acid. **b** Quantum efficiency of copper colloid doped devices. Reproduced with permission [10]. Copyright 2020, Wiley

in which strong p-doping effect, antireflection effect and an encapsulation effect could be confirmed simultaneously. Interestingly, all of three photovoltaic parameters (J_{SC}, V_{OC} and FF) increase with the enhancement of light absorption, and the improved conductivity and the Fermi level downshift of SWCNTs. As a result, the Nafion-doped SWCNT/Si solar cells exhibited an enhanced PCE from 9.5 to 14.4%. Owing to the superior encapsulation effect of Nafion film, the device shows the highly stabile PCE more than 120 days under severe conditions. Tune et al. [10] further developed a high-efficiency SWCNT/Si solar cell by optimizing Nafion-doped SWCNT/DWCNTs, contact finger geometry and random pyramid texture etch, in which high conductivity and high transparency Nafion-doped SWCNT/DWCNT networks contacted with pyramid textured Si (Fig. 7.9). After the process and design optimization, the efficiency was broken through, yielding the PCEs of 17.2 and 15.5% for the active areas of 1 and 5 cm^2, respectively. Furthermore, Chen et al. [23] demonstrated the high-efficiency large-area (1–16 cm^2) SWCNT/Si solar cells with industry standard device geometries by a front and back-junction design, in which intermixed Nafion/SWCNT layers were used to form the vdW heterojunction with Si at both front and back side. As a result, the new record PCEs of 18.9% had been achieved for the front and back-junction solar cell.

7.4.3 Improving the Hole Transportation of SWCNT Films

In order to improve the hole transport, extra transport materials have usually been introduced into the SWCNT/Si vdW heterojunction solar cells. For example, polyaniline as an interlayer has been demonstrated to enhance the performance of SWCNT/Si solar cells by increasing the physical separation between photogenerated holes and electrons, resulting in a PCE of 9.66% after chemical treatments

[47]. Yu et al. [29] has added a solid-state hole transporting material (spiro-OMeTAD) as an interlayer between GO-SWCNT and Si to fabricate the GO-SWCNT/spiro-OMeTAD/Si solar cells. Due to suitable bandgap structure relative to Si, the spiro-OMeTAD interlayer could be able to significantly minimize the recombination at the interface between GO-SWCNT and Si. Consequently, the PCE of devices could be improved to 9.49% from 6.80% by applying a spiro-OMeTAD layer, and a high PCE of 12.83% was obtained after optimizing the interlayer thickness, the transmittance of the GO-SWCNT film and $AuCl_3$ doping.

In addition, inorganic interlayers have also be used as hole transport materials due to high hole mobility and good stability. For example, black phosphorus (BP) sheets was also found to enhance the performance of SWCNT/Si heterojunction solar cells by incorporating BP nanosheets into SWCNTs [28]. The PCE of 9.37% could be achieved due to improved charge transfer properties and suppressed recombination rate. Wang et al. [48] demonstrated that MoO_x as efficient hole transport layer could minimize the interfacial recombination and reduce the height of the Schottky barrier at the SWCNT/Au interface in a SWCNT/Si heterojunction solar cell, resulting in a lower reverse leakage current and higher rectification ratio (Fig. 7.10). Moreover, the MoO_x layer also served as an antireflection layer for the solar light and as a carrier dopant for the SWCNTs. The PCE of SWCNT/Si heterojunction solar cell increased from 11.1 to 17.0% at the optimal conditions.

7.4.4 Introducing Insulating Interfacial Layers

As we know, the static charge transfer often occurs when SWCNTs contact directly with Si substrate, resulting in a decrease of built-in electric field at the interface. On the other hand, the recombination of photogenerated carrier usually happens at the interface due to a specific charge distribution and the interfacial defect states. Therefore, appropriate interface layers have been introduced to passivate the interfacial defects and suppress the static charge transfer, thereby improving the PCE of SWCNT/Si vdW heterojunction solar cells. Under the ambient conditions, a thin native oxide layer are easily formed at the SWCNT/n-type Si interface, which will affect the PCE of the solar cell if the thickness further increases. For a pristine SWCNT/Si solar cell, the optimum thickness of native oxide layer is believed to be $\sim$0.7 nm [9, 49]. Due to the existence of the insulating oxide layer, the barrier height of SWCNT/Si heterojunction is enlarged, and the transport form of carriers through the insulating layer would either be tunneling or recombination. However, a thicker oxide layer will block the carrier transport at the interface, resulting in a degraded photovoltaic performance, this is why the pristine SWCNT/Si solar cells are not stable after exposure to air for a certain days.

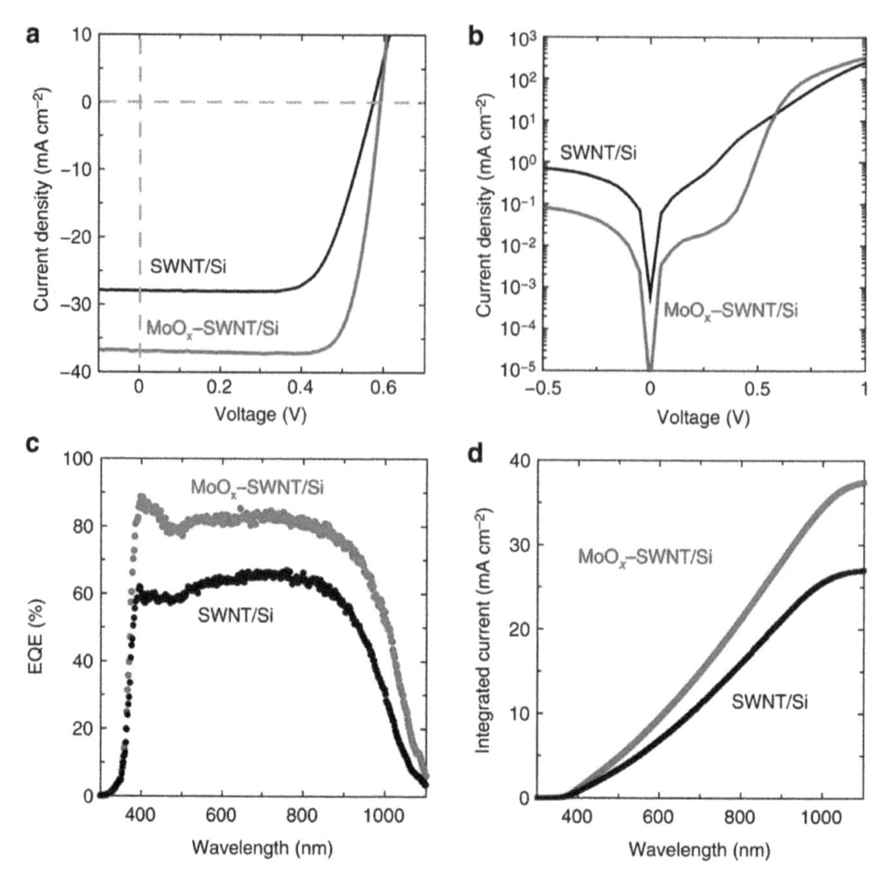

Fig. 7.10 **a** J-V curves and **b** log-plot of the dark current for an SWCNT/Si solar cell recorded under AM 1.5 G conditions before and after the deposition of an MoO_x layer. **c** IPCE spectra and **d** calculated integrated photocurrent under AM 1.5 G irradiation before and after the deposition of the MoO_x coating. Reproduced with permission [48]. Copyright 2018, Nature

7.5 SWCNT/GaAs vdW Heterojunction Solar Cells

Except for Si, GaAs is another important semiconductor with a direct bandgap, high electron mobility and large optical absorption coefficient ($\sim 10^4$/cm), which has been widely used as highly efficient space solar cells [50]. Therefore, the heterojunctions composed of SWCNT and GaAs have a greater potential to exhibit higher photovoltaic performance. Liang et al. [51] firstly investigated the transport properties of a SWCNT/GaAs vdW heterojunction by transferring an individual SWCNT and on the top of GaAs substrate. The rectifying behavior of the SWCNT/n-type GaAs vdW heterojunction has been demonstrated while the ohmic contact forms between SWCNT and p-doped GaAs (Fig. 7.11a). Moreover, an obvious photovoltaic effect in the SWCNT/n-GaAs vdW heterojunction has been confirmed

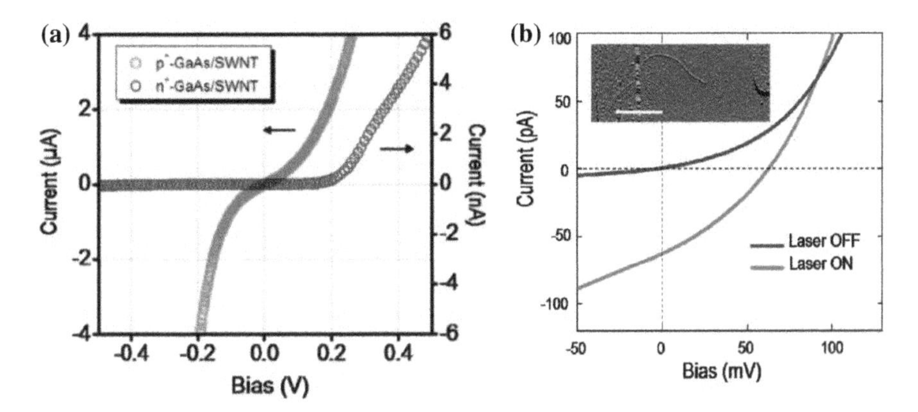

Fig. 7.11 **a** I-V curves of SWCNT/n- and p-GaAs vdW heterojunction devices. **b** Typical I-V curves with and without light illumination. Reproduced with permission [51]. Copyright 2008, American Chemical Society

under the light illumination, demonstrating the feasibility of the SWCNT/n-GaAs vdW heterojunction solar cells (Fig. 7.11b). The conduction mechanisms of p/n heterojunctions have also been studied. Li et al. [52] further analyzed the detailed carrier transport model of sc-SWCNT/n-type GaAs heterojunction solar cells. At a low and high-forward bias, the electron transports have been demonstrated to be dominated by the thermionic emission and tunneling processes, respectively. However, at low and high-reverse bias, the direct and Fowler-Nordheim tunneling were considered to be the dominant transport mechanisms, respectively.

In the vdW heterostructures, sc-SWCNTs are not only used to separate and transport the photogenerated carriers in GaAs, but also act as the light absorber to generate carriers under light illuminations. Thus, the interface states between sc-SWCNTs and GaAs determines the photoelectric properties of the SWCNT/n-GaAs vdW heterojunction solar cells. The interface recombination between SWCNT and GaAs have an important effect on the J_{sc} and V_{oc}. Especially, the increasing of interface recombination results in a remarkable decrease of the V_{oc} value, and the J_{sc} value is also reduced to some extent [53, 54]. In addition, the SWCNT as carrier transporting layer has also been demonstrated to enhance the collection efficiency of photocurrent via optimization of the optical transmittance and electronic transport properties [55].

Recently, our group developed a highly efficient photovoltaic devices based on (6, 5)-enriched sc-SWCNT/GaAs vdW heterojunctions. By replacing the mixed SWCNT film with (6, 5)-enriched sc-SWCNT film, a typical II-type p–n junction has been formed at the interface due to the suitable bandgap alignment. After optimizing the transfer process of (6, 5)-enriched sc-SWCNT film, the vdW heterojunction solar cells have been successfully fabricated, as shown in Fig. 7.12a. The dark J-V curve shows an excellent rectification characteristic, suggesting the formation of vdW heterojunctions between the (6, 5)-enriched sc-SWCNT film and

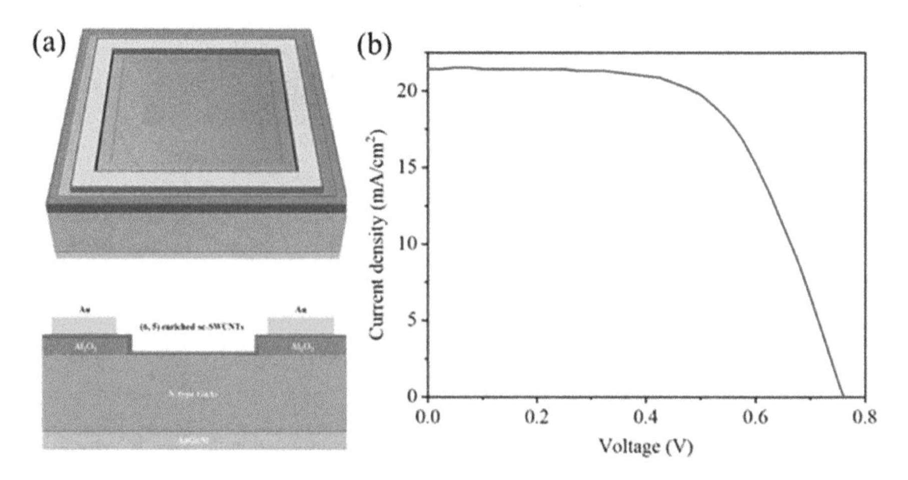

Fig. 7.12 **a** Diagram schematic and optical photo of sc-SWCNT/n-GaAs solar cells. **b** A typical J-V curve under AM 1.5 G illuminations

n-type GaAs. Under AM 1.5 G light illumination (100 mW/cm^2), an apparent photovoltaic effect can be observed from the J-V curve (Fig. 7.12b), a new record PCE of 10.15% has been first achieved with a J_{sc} of 21.55 mA/cm^2, V_{oc} of 0.76 V and FF of 0.62. Although the PCE is still lower than those of SWCNT/Si solar cells, the separation and transportation of photogenerated carriers can be enhanced by optimizing the interface contact and p-doping (6, 5)-enriched sc-SWCNTs.

7.6 Summary and Outlook

The unique structure and excellent electric and optical properties make CNTs have obvious advantages for photovoltaic applications, and highly efficient vdW heterojunctions solar cells have been demonstrated by combining with bulk semiconductor due to the simple device structure and easy fabrication. We mainly introduce the state-of-the-art research progress about the solar cells based on the CNT/semiconductor vdW heterojunctions, including DWCNT or MWCNT/Si, SWCNT/Si and SWCNT/GaAs. Especially, different techniques have been optimized to improve the photovoltaic performance of CNT/Si vdW heterojunction solar cells, such as high quality CNT film, Fermi level controlling, and interface engineering, and so on. The recent PCEs of DWCNT or MWCNT/Si, SWCNT/Si solar cell have been improved to over 15%, implying a great potential to fabricate highly efficient vdW heterojunction solar cells. However, the PCE is still far lower than those of commercial Si solar cells, more efforts are still required to further optimize the properties of vdW heterojunctions and interface engineering. Meanwhile, the solar cells base on SWCNT/GaAs heterojunctions should be paid more attention, since the device structure and working mechanism are the same.

The optimized SWCNT/GaAs vdW heterojunction solar cells theoretically higher PCE then SWCNT/Si heterojunctions due to the remarkable advantage of GaAs. Once the related issues can be well solved, the SWCNT/semiconductor vdW heterojunctions will be very promising for the fabrication of next-generation highly efficient photovoltaic cells.

However, before reaching the PCE of commercial solar cells, there are still many efforts to be made in the future work, especially the following three aspects should be focused on:

Firstly, the basic physics of SWCNT/semiconductor vdW heterojunctions are still needed to be further understood. Especially, the detailed charge transfer dynamic between SWCNT and bulk semiconductor has not be investigated systematically under light illumination due to unique electronic structures of SWCNTs. Meanwhile, the theoretical design based on the device mode and the SWCNTs with suitable chirality can give a guide to the experimental fabrication of SWCNT/semiconductor vdW heterojunction solar cells.

Secondly, the optimization of CNT film is the most important part in the CNT/semiconductor vdW heterojunction solar cells. To a large extent, the physical or chemical states of CNTs determines the properties of CNT film and the related photovoltaic performance, including the nanotube types, chirality/diameter, lengths, surface cleanliness, chemical doping, film morphology, and so on. Especially, high recombination probability is easily observed at the sc-SWCNT film with mixed chirality, which will hinder the highly efficient charge transport and reduce the collection of photocurrents. In addition, since the main role of CNT film is to separate and transport the photogenerated carriers, high quality SWCNTs are more suitable than DWCNTs or MWCNTs due to higher transparency.

Thirdly, the surface state of bulk semiconductor (Si or GaAs) also have an important effect on the photovoltaic performance of vdW heterojunction solar cells, there are a large number carrier combination sites at the surface of semiconductors. Thereby, the surface passivation techniques are still further explored and optimized, especially for GaAs substrate. In addition, the anti-reflection layer, light management and the ohmic contacts between SWCNT and electrode should also be deeply investigated.

References

1. Green MA, Dunlop ED, Hohl-Ebinger J, Yoshita M, Kopidakis N, Hao X (2020). Solar cell efficiency tables (version 56). Prog Photovoltaics Res Appl, 28(7): 629–638.
2. Hu L, Hecht DS, Gruner G (2010). Carbon nanotube thin films: fabrication, properties, and applications. Chem Rev, 110(10), 5790–5844.
3. Cai B, Su Y, Tao Z, Hu J, Zou C, Yang Z, Zhang Y (2018). Highly sensitive broadband single-walled carbon nanotube photodetectors enhanced by separated graphene nanosheets. Adv Optical Mater, 6(23): 1800791.

4. Li XM, Lv Z, Zhu HW (2015). Carbon/silicon heterojunction solar cells: state of the art and prospects. Adv Mater 27: 6549–6574.
5. Classen A, Einsiedler L, Heumueller T, Graf A, Brohmann M, Berger F, Kahmann S, Richter M, Matt GJ, Forberich K, Zaumseil J, Brabec, CJ (2019). Absence of charge transfer state enables very low VOC losses in SWCNT: fullerene solar cells. Adv Energy Mater, 9(1), 1801913.
6. Habisreutinger SN, Nicholas RJ, Snaith HJ (2017). Carbon nanotubes in perovskite solar cells. Adv Energy Mater, 7(10), 1601839.
7. Wei J, Jia Y, Shu Q, Gu Z, Wang K, Zhuang D, Zhang G, Wang Z, Luo J, Cao A, Wu D (2007). Double-walled carbon nanotube solar cells. Nano Lett 7(8): 2317–2321.
8. Tune DD, Flavel BS (2018). Advances in carbon nanotube–silicon heterojunction solar cells. Adv Energy Mater, 8(15): 1703241.
9. K Cui, AS Anisimov, T Chiba, S Fujii, H Kataura, AG Nasibulin, S Chiashi, EI Kauppinen, S Maruyama (2014). Air-stable high-efficiency solar cells with dry-transferred single-walled carbon nanotube films. J Mater Chem A, 2: 11311–11318.
10. Tune DD, Mallik N, Fornasier H, Flavel BS (2020). Breakthrough carbon nanotube-silicon heterojunction solar cells. Adv Energy Mater, 10(1): 1903261.
11. Xu W, Wu S, Li X, Zou M, Yang L, Zhang Z, Wei J, Hu S, Li Y, Cao A (2016). High-efficiency large-area carbon nanotube-silicon solar cells. Adv Energy Mater, 6(12): 1600095.
12. Zhao X, Wu H, Yang L, Wu Y, Sun Y, Shang Y, Cao A (2019). High efficiency CNT-Si heterojunction solar cells by dry gas doping. Carbon, 147: 164–171.
13. Jia Y, Wei J, Wang K, Cao A, Shu Q, Gui X, Zhu Y, Zhuang D, Zhang G, Ma B, Wang L (2008). Nanotube-silicon heterojunction solar cells. Adv Mater, 20(23): 4594–4598.
14. Jia Y, Li PX, Gui XC, Wei JQ, Wang KL, Zhu HW, Wu DH, Zhang LH, Cao AY, Xu Y (2011). Encapsulated carbon nanotube-oxide-silicon solar cells with stable 10% efficiency. Appl Phys Lett, 98: 133115.
15. Di JT, Yong ZZ, Zheng XH, Sun BQ, Li QW (2013). Aligned carbon nanotubes for high-efficiency Schottky solar cells. Small, 9(8): 1367–1372.
16. X Gan, RT Lv, JF Bai, ZX Zhang, JQ Wei, ZH Huang, HW Zhu, FY Kang, M Terrones (2015). Efficient photovoltaic conversion of graphene–carbon nanotube hybrid films grown from solid precursors. 2D Mater, 2: 034003.
17. Fan Q, Zhang Q, Zhou W, Xia X, Yang F, Zhang N, Xiao S, Li K, Gu X, Xiao Z, Chen H, Wang Y, Liu H, Zhou W, Xie S (2017). Novel approach to enhance efficiency of hybrid silicon-based solar cells via synergistic effects of polymer and carbon nanotube composite film. Nano Energy, 33: 436–444.
18. Wu H, Zhao X, Sun Y, Yang L, Zou M, Zhang H, Wu Y, Dai L, Shang Y, Cao A (2019). Improving carbon nanotube-silicon solar cells by solution processable metal chlorides. Solar RRL, 3(8): 1900147.
19. Wu H, Zhao X, Wu Y, Ji Q, Dai L, Shang Y, Cao A (2020). Improving CNT-Si solar cells by metal chloride-to-oxide transformation. Nano Res, 13(2): 543–550.
20. E Shi, L Zhang, Z Li, P Li, Y Shang, Y Jia, J Wei, K Wang, H Zhu, D Wu, S Zhang, A Cao (2012). TiO_2-coated carbon nanotube-silicon solar cells with efficiency of 15%. Sci Rep, 2: 884.
21. Li ZR, Kunets VP, Saini V, Xu Y, Dervishi E, Salamo GJ, Biris AR, Biris AS (2009). Light-harvesting using high density p-type single wall carbon nanotube/n-type silicon heterojunctions. ACS Nano, 3(6): 1407–1414.
22. Harris JM, Headrick RJ, Semler MR, Fagan JA, Pasquali M, Hobbie EK (2016). Impact of SWCNT processing on nanotube-silicon heterojunctions. Nanoscale, 8(15): 7969–7977.
23. Chen J, Tune DD, Ge K, Li H, Flavel BS (2020). Front and back-junction carbon nanotube-silicon solar cells with an industrial architecture. Adv Funct Mater, 30(17): 2000484.
24. Jung Y, Li XK, Rajan NK, Taylor AD, Reed MA (2013). Record high efficiency single-walled carbon nanotube/silicon p-n junction solar cells. Nano Lett, 13: 95–99.

25. Cui KH, Chiba T, Omiya S, Thurakitseree T, Zhao P, Fujii S, Kataura H, Einarsson E, Chiashi S, Maruyama S (2013). Self-assembled microhoneycomb network of single-walled carbon nanotubes for solar cells. J Phys Chem, Lett 4: 2571–2576.
26. Wang F, Kozawa D, Miyauchi Y, Hiraoka K, Mouri, S Ohno Y, Matsuda K (2014). Fabrication of single-walled carbon nanotube/Si heterojunction solar cells with high photovoltaic performance. ACS Photonics, 1: 360–364.
27. Li X, Jung Y, Sakimoto K, Goh TH, Reed MA, Taylor AD (2013). Improved efficiency of smooth and aligned single walled carbon nanotube/silicon hybrid solar cells. Energy Environ Sci, 6: 879–887.
28. Bat-Erdene M, Batmunkh M, Tawfik SA, Fronzi M, Ford MJ, Shearer CJ, Yu L, Dadkhah M, Gascooke JR, Gibson CT, Shapter JG (2017). Efficiency enhancement of single-walled carbon nanotube-silicon heterojunction solar cells using microwave-exfoliated few-layer black phosphorus. Adv Funct Mater, 27(48): 1704488.
29. Yu L, Batmunkh M, Grace T, Dadkhah M, Shearer C, Shapter J (2017). Application of a hole transporting organic interlayer in graphene oxide/single walled carbon nanotube-silicon heterojunction solar cells. J Mater Chem A, 5(18): 8624–8634.
30. Alekseeva AA, Rajanna PM, Anisimov AS, Sergeev O, Bereznev S, Nasibulin AG (2018). Synergistic effect of single-walled carbon nanotubes and PEDOT:PSS in thin film amorphous silicon hybrid colar cell. Phys Status Solidi B, 255: 1700557.
31. Hu XG, Hou PX, Liu C, Zhang F, Liu G, Cheng HM (2018). Small-bundle single-wall carbon nanotubes for high-efficiency silicon heterojunction solar cells. Nano Energy, 50: 521–527.
32. Hu XG, Hou PX, Wu JB, Li X, Luan J, Liu C, Liu G, Cheng HM (2020). High-efficiency and stable silicon heterojunction solar cells with lightly fluorinated single-wall carbon nanotube films. Nano Energy, 69: 104442.
33. Alzahly S, Yu L, Shearer CJ, Gibson CT, Shapter JG (2018). Efficiency improvement using molybdenum disulphide interlayers in single-wall carbon nanotube/silicon solar cells. Mater, 11(4): 639.
34. Tune DD, Hennrich F, Dehm S, Klein MF, Glaser K, Colsmann A, Shapter JG, Lemmer U, Kappes MM, Krupke R, Flavel BS (2013). The role of nanotubes in carbon nanotube-silicon solar cells. Adv Energy Mater, 3(8): 1091–1097.
35. Z. Wu, Z. Chen, X. Du, J.M. Logan, J. Sippel, M. Nikolou, K. Kamaras, J.R. Reynolds, D.B. Tanner, A.F. Hebard (2004). Transparent, conductive carbon nanotube films. Science, 305 (5688), 1273–1276.
36. Shi Z, Chen X, Wang X, Zhang T, Jin J (2011). Fabrication of superstrong ultrathin free-standing single-walled carbon nanotube films via a wet process. Adv Funct Mater, 21 (22), 4358–4363.
37. Gilshteyn EP, Romanov SA, Kopylova DS, Savostyanov GV, Anisimov AS, Glukhova OE, Nasibulin AG (2019). Mechanically tunable single-walled carbon nanotube films as a universal material for transparent and stretchable electronics. ACS Appl. Mater. Interfaces, 11 (30), 27327–27334.
38. Walker JS, Fagan JA, Biacchi AJ, Kuehl VA, Searles TA, Hight Walker AR, Rice WD (2019). Global alignment of solution-based single-wall carbon nanotube films via machine-vision controlled filtration. Nano Lett, 2019, 19(10): 7256–7264.
39. Kaskela A, Nasibulin AG, Timmermans MY, Aitchison B, Papadimitratos A, Tian Y, Zhu Z, Jiang H, Brown DP, Zakhidov A, Kauppinen EI (2010). Aerosol-synthesized SWCNT networks with tunable conductivity and transparency by a dry transfer technique. Nano Lett, 10(11), 4349–4355.
40. Tsapenko AP, Goldt AE, Shulga E, Popov ZI, Maslakov KI, Anisimov AS, Sorokin PB, Nasibulin AG (2018). Highly conductive and transparent films of $HAuCl_4$-doped single-walled carbon nanotubes for flexible applications. Carbon, 130, 448–457.
41. Jiang S, Hou PX, Liu C, Cheng HM (2019). High-performance single-wall carbon nanotube transparent conductive films. J Mater Sci Tech, 35(11), 2447–2462.
42. Harris JM, Semler MR, May S, Fagan JA, Hobbie EK (2015). Nature of record efficiency fluid-processed nanotube-silicon heterojunctions. J Phys Chem C, 119(19): 10295–10303.

43. Li XK, Jung Y, Huang JS, Goh T, Taylor AD (2014). Device area scale-up and improvement of SWNT/Si solar cells using silver nanowires. Adv Energy Mater, 4, 1400186.
44. Jia Y, Cao AY, Bai X, Li Z, Zhang LH, Guo N, Wei JQ, Wang KL, Zhu HW, Wu DH, Ajayan PM (2011). Achieving high efficiency silicon-carbon nanotube heterojunction solar cells by acid doping. Nano Lett, 11(5): 1901–1905.
45. Cui K, Qian Y, Jeon I, Anisimov A, Matsuo Y, Kauppinen EI, Maruyama S (2017). Scalable and solid-state redox functionalization of transparent single-walled carbon nanotube films for highly efficient and stable solar cells. Adv Energy Mater, 7(18): 1700449.
46. Qian Y, Jeon I, Ho YL, Lee C, Jeong S, Delacou C, Seo S, Anisimov A, Kaupinnen EI, Matsuo Y, Kang Y (2020). Multifunctional effect of p-doping, antireflection, and encapsulation by polymeric acid for high efficiency and stable carbon nanotube-based silicon solar cells. Adv Energy Mater, 10(1): 1902389.
47. Tune DD, Flavel BS, Quinton JS, Ellis AV, Shapter JG (2013). Single-walled carbon nanotube/polyaniline/n-silicon solar cells: Fabrication, characterization, and performance measurements. ChemSusChem, 6(2): 320–327.
48. Wang F, Kozawa D, Miyauchi Y, Hiraoka K, Mouri S, Ohno Y, Matsuda K (2015). Considerably improved photovoltaic performance of carbon nanotube-based solar cells using metal oxide layers. Nat Commun, 6(1), 6305.
49. Hu X, Hou PX, Liu C, Cheng HM (2019). Carbon nanotube/silicon heterojunctions for photovoltaic applications. Nano Mater Sci, 1(3): 156–172.
50. Huo TT, Yin H, Zhou DY, Sun LJ, Tian T, Wei H, Hu NT, Yang Z, Zhang YF, Su YJ (2020). A self-powered broadband photodetector based on single-walled carbon nanotube/GaAs heterojunctions. ACS Sustainable Chem Eng, 8(41): 15532–15539.
51. Liang CW, Roth S (2008). Electrical and optical transport of GaAs/carbon nanotube heterojunctions. Nano Lett, 8(7): 1809-1812.
52. Li H, Loke WK, Zhang Q, Yoon SF (2010). Physical device modeling of carbon nanotube/GaAs photovoltaic cells. Appl Phys Lett, 96 (4): 043501.
53. Khadije K, Asgari A, Movla H, Mottaghizadeh A, Najafabadi HA (2011). Effects of interface recombination on the performance of SWCNT\GaAs heterojunction solar cell. Procedia Eng, 8: 275–279.
54. Movla H, Ghaffari S, Rezaei E (2016). A numerical study on the influence of interface recombination on performance of carbon nanotube/GaAs solar cells. Opt Quant Electron, 48 (8): 390.
55. Mitin DM, Bolshakov AD, Neplokh V, Mozharov AM, Raudik SA, Fedorov VV, Shugurov KY, Mikhailovskii VY, Rajanna PM, Fedorov FS, Nasibulin AG, Mukhin IS (2020). Novel design strategy for GaAs-based solar cell by application of single-walled carbon nanotubes topmost layer. Energy Sci Eng, 8(8), 2938–2945.

Chapter 8
Toward All-Carbon Hybrid Solar Cells

8.1 Introduction

The sp^n hybridizations (n = 1 ~ 3) of carbon atoms enable various nanomaterials with zero-, one- and two-dimensional structures, such as fullerenes, carbon nanotubes (CNTs), graphene, and so on. The physical and chemical properties strongly depend on their hybridizations and atomic arrangement [1, 2]. For instance, 0D fullerenes are typical n-type semiconductors while 1D single-walled CNTs (SWCNTs) can be either metallic or semiconducting depending on their diameter and chirality. 2D Graphene exhibits ultrahigh carrier mobility and thermal conductivity at room temperature, which is a thinnest transparent conducting material used in optoelectronic and electronic devices [3, 4]. Therefore, all-carbon nanodevices are expected to be fabricated using these carbon nanomaterials as building blocks since these carbon allotropes behave as semiconductors or conductors depending on the atomic arrangements [5].

Semiconducting SWCNTs (sc-SWCNTs) exhibit unique and outstanding optoelectronic properties at room temperature, such as strong optical absorptivity ($\alpha > 10^5$ cm^{-1}), diameter-depended bandgaps, ultrahigh carrier mobility, solution-processability, and excellent chemical stability. Especially, strong optical absorptivity enables sc-SWCNTs effectively collect the incident lights in films as thin as 100 nm, and the bandgaps of sc-SWCNTs can change from 0.5 to 1.2 eV when the diameters range from 0.7 to 2.0 nm, [6] which makes sc-SWCNTs be promising candidates as the active layer for next-generation photovoltaic devices. However, the effective exciton dissociation in the sc-SWCNTs needs to be completed before transfer and transportation. As we know, the enhanced Coulomb interaction in the confined 1D sc-SWCNT results in the formation of excitons with diameter-depended binding energy under illuminations, which sometimes dominates the optical absorption of sc-SWCNTs [7]. Generally, the dissociation of excitons in the sc-SWCNTs is not efficient without an external driving force, thereby well-designed heterojunctions are usually needed to enhance the

Y. Su, *High-Performance Carbon-Based Optoelectronic Nanodevices*,
Springer Series in Materials Science 319,
https://doi.org/10.1007/978-981-16-5497-8_8

dissociation. In addition, the interband transitions (S_{11}, S_{22} and S_{33}) between van Hove singularities determine the optical absorption and corresponding photoelectric properties of sc-SWCNTs. It has been demonstrated that the sc-SWCNT based photovoltaic devices mainly response to NIR light due to the larger absorption cross section of the S_{11} transition than those of S_{22} and S_{33}. Although the optical absorption induced by S_{22} and S_{33} transitions also contributes to photocurrents, more than 40% light absorption is due to the S_{11} transition for a 7 nm sc-SWCNT film [8].

In this chapter, we mainly introduce the photoexcitation transfer dynamics and bandgap structure limits in the sc-SWCNT/fullerene heterojunctions. And then, the research progress on carbon-based solar cells have been summarized, where sc-SWCNT/fullerene bulk or planar heterojunctions are used as active layers. Finally, the concept, research progress and current challenges about all-carbon solar cells have been introduced. And the future perspectives of all-carbon solar cells have also been highlighted.

8.2 Excitation Transfer Dynamics of Sc-SWCNTs and Sc-SWCNT/Fullerenes

As we know, the separation and recombination of photogenerated carriers happen within a very short time, the time scales of separation and transfer process will determine the photoelectric efficiency of photovoltaic devices. Therefore, the better understanding of photoexcitation transfer dynamics in sc-SWCNT film is essential for the design of carbon-based photovoltaic devices. It is reported that only a part of the excitons can be transferred over 5–10 nm into nearby nanotube bundle within first 200–400 fs of optical excitation, [9] Most of them can travel large distances aligned tubes on a ~ 10 ps time scale before recombination and trapping, and the charge hop between bundles occur only at the intersections [10]. For the hot excitons generated by high-energy photons, most of their excess energy is loosed within the first 100 fs after photoexcitation, [11] which make the generation of multiple excitons be negligible owing to the enhanced electronic screening and ultrafast intertube exciton transport in sc-SWCNT films. Consequently, the heterojunction interface is critically needed to facilitate ultrafast exciton dissociation in sc-SWCNT based solar cell devices, which outcompete the rapid decay of hot excitons and exciton trapping in sc-SWCNT thin films [11]. Meanwhile, the thickness of sc-SWCNT films is also need to be well designed so that the excitations after hoping between bundles reach the interface of heterojunctions, as well as the thickness-dependent free carrier yield[12].

After combining sc-SWCNTs with the electron-accepting C_{60} layer, ultrafast exciton dissociation will happen, and the photoinduced electron transfer time from sc-SWCNT to C_{60} is evaluated to be ≤ 120 fs with a yield of $38 \pm 3\%$ [13]. The lifetime of free carriers generated in polymer-wrapped sc-SWCNT films can be much larger than 100 ns due to the charge separation at interface, and the corresponding driving force shows diameter-dependent feature and vanishes for large-diameter sc-SWCNTs [14]. In the C_{60} layer, the diffusion time of photogenerated exciton to the interface is over hundreds of picoseconds, implying that a singlet exciton diffusion length in C_{60} layer is approximately 5 nm [15].

8.3 Bandgap Structure Limits of SWCNT/Fullerenes Hybrids

Although the photogenerated electron transfer from sc-SWCNT to C_{60} can finished within 120 fs at the interface, not all electron transfer process happens in the sc-SWCNT/fullerene hybrids. It has been demonstrated that the electron transfer dynamics depends on the bandgap of sc-SWCNTs and the energy offset between sc-SWCNT to C_{60}. The (n, m) chiral index of sc-SWCNTs directly determine their diameter and the magnitude of the optical transitions. For most of sc-SWCNTs, the main S_{11} optical transition absorption occurs in the near-infrared region. While the S_{22} optical transition can be observed by absorbing visible (or NIR) lights in the sc-SWCNTs with small (large) diameter. The weak light absorption induced by the S_{22} optical transition is also found in the UV or visible region from the UV-vis-NIR spectra due to the diameter difference. Therefore, a close match between sc-SWCNTs and solar spectrum can be obtained theoretically by combing different sc-SWCNTs with (n, m) species. Tune et al. [16] predicted that a sunlight harvesting potential of $\sim$28% (or $\sim$19%) could be obtained for the idealized tandem solar cell consisting of four small- (or large) diameter sc-SWCNTs. However, owing to the diameter-related binding energy of excites in sc-SWCNTs, the dissociation of excitons in sc-SWCNTs is usually achieved by the formation of type-II heterojunction using fullerenes as electron acceptors, in which a minimum energy offset at the heterointerface exceed excitons binding energy. Therefore, only a few sc-SWCNTs with certain bandgaps can be used to fabricate all-carbon solar cells with specific fullerene-based acceptors.

Pfohl et al. [8] systematically investigated the diameter limit of surfactant wrapped sc-SWCNTs and the required net driving energy for sc-SWCNT exciton dissociation in the SWCNT/fullerene solar cells by changing the chirality of sc-SWCNTs. As shown in Fig. 8.1, an upper limit of diameters in the polymer-free SWCNT/C_{60} solar cells was found to be 0.95 nm by considering the certain band offset between the LUMO levels of sc-SWCNTs and fullerenes. Moreover, no photocurrent was considered to be generated through the S_{22} optical transition of nanotubes with a larger diameter due to faster energy relaxation from S_{22} to S_{11}. When the sc-SWCNTs with different diameters were used to fabricate the SWCNT/fullerene solar cells, the potential matching fullerenes were proposed to create a required net driving energy at the junctions. For the more negative LUMO of fullerenes, the sc-SWCNTs with larger diameter could be used to fabricate

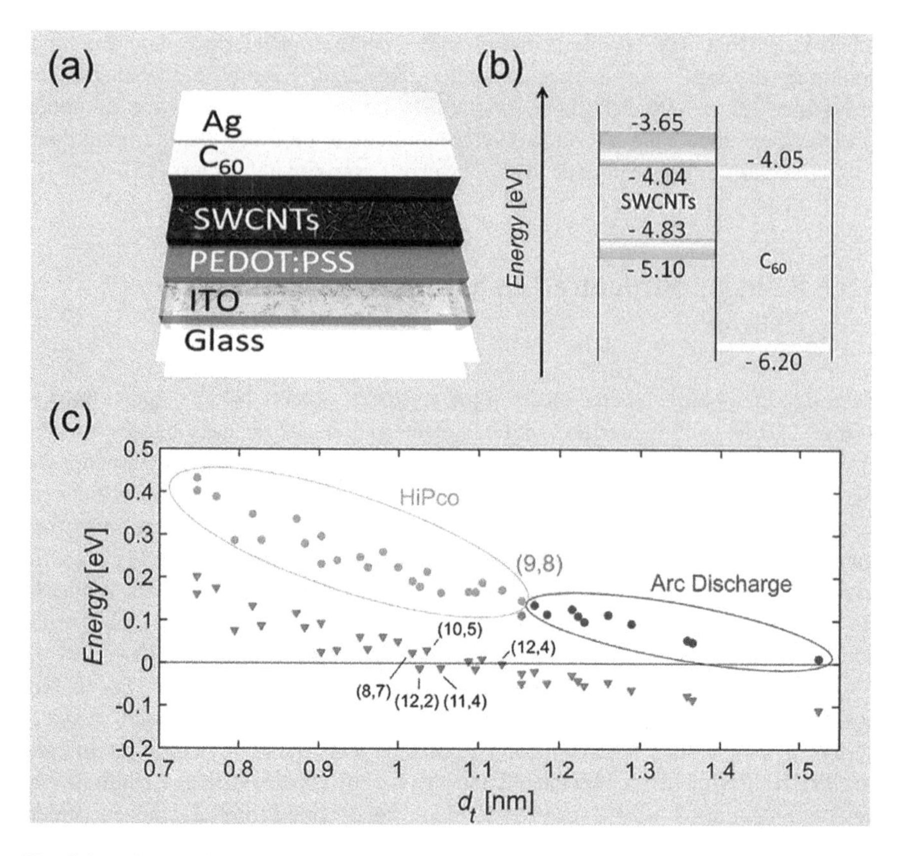

Fig. 8.1 **a** Schematic of the solar cell architecture based on the SWCNT/fullerene hybrids. **b** Energy diagram of SWCNTs with diameter between 0.7 and 1.8 nm. **c** Net driving energy for exciton dissociation for SWCNTs with different diameters. Reproduced with permission [8]. Copyright 2016, Wiley

all-carbon hybrid solar cells. In addition, the difference of upper diameter limit between polymer-free and polymer wrapped nanotubes was also discussed by considering the differences in dielectric environment.

8.4 All-Carbon Nanohybrids as Active Layer

In the past ten years, the photovoltaic devices based on sc-SWCNTs have attracted increasing attention due to ultrahigh carrier mobility, unique electron structures, high light absorption coefficient, and high chemical stability of sc-SWCNTs, as well as low-cost processing. Especially, the sc-SWCNTs are promising candidates as the active layer in NIR photovoltaic devices, where the excitons are generated by E_{11} transition of sc-SWCNTs. The photogenerated excitons can be effectively dissociated and separated into free charge carriers using fullerenes as electron acceptors, forming a photocurrent and photovoltage in an external circuit. Generally, the as-formed sc-SWCNT/fullerene heterojunctions as photoactive layer can be divided into two types: bulk and planar heterojunctions according to the contact type and interfacial area.

8.4.1 All-Carbon Bulk Heterojunctions

Bindl et al. [17] demonstrated a solution processed photovoltaic solar cells NIR photovoltaic device based on carbon bulk heterojunctions (BHJ) between sc-SWCNTs and $PC_{61}BM$, in which photogenerated charges could be efficiently separated by the favorable energy offsets and enhanced by the ultrahigh interfacial area without exciton diffusion limitations. Owing to the enhanced absorptivity in the NIR by the sc-SWCNTs, the device exhibited a NIR PCE of 1.4% and max external QE of 18.3% at $\lambda = 1205$ nm. Gong et al. [18] further maximized the photocurrent of SWCNT/$PC_{71}BM$ BHJ solar cells with inverted device structures using polychiral sc-SWCNTs, in which ZnO nanowires as an electron transport layer interpenetrated into the carbon active layer to minimize the collection length. Importantly, both light absorptions from fullerenes and sc-SWCNTs were contributed to high photocurrent in the inverted cell due to the improved charge extraction from the active layer. Consequently, the device exhibited average NREL certified and max PCEs of 2.48% and 3.1% without reducing the Voc, respectively. By passivizing the surface traps of ZnO nanowires using polyethylenimine, the PCE of carbon-based BHJ solar cells increased to be 3.2% under AM 1.5 1 sun illumination [19]. However, with increasing the active area of carbon-based solar cells, the overall performance reduced significantly due to the spatial inhomogeneity in the sc-SWCNT/fullerene film. Shastry et al. [20] demonstrated that 1,8-diiodooctane (DIO) as a solvent additive could improve the morphologies of sc-SWCNT/$PC_{71}BM$ BHJ film by reducing the size of fullerene aggregates, large-areas (6–50 mm^2) carbon-based BHJ solar cells have been fabricated with excellent performance that is comparable to smaller area devices. By optimizing the DIO processing conditions and the chiral distribution of enriched (7, 6) sc-SWCNTs, a large-area (6 mm^2) device has been demonstrated to exhibit a champion efficiency of 2.31%.

Bernardi et al. [21] demonstrated a solution-processable BHJ solar cell with a PCE of 1.3% using carbon active layers composed of $PC_{70}BM$ fullerene, sc-SWCNTs and reduced graphene oxide (rGO), which exhibited a highest Jsc of 3.1 mA/cm^2, a Voc of 0.75 V and a FF of 0.55. A type-II alignment between sc-SWCNTs and $PC_{70}BM$ was formed in the carbon-based BHJ solar cell, in which sc-SWCNTs worked as the donor and the [6]-phenyl-C61-butyric acid methyl ester (PCBM) worked as the acceptor. It was found that most of the photocurrent and quantum efficiency was contributed by $PC_{70}BM$ as the main active constituent, and the efficiency limits were calculated to be 9% and 13% for sc-SWCNTs with the diameter of 1.2–1.7 nm and 0.75–1.2 nm, respectively. Isborn et al. [22] further evaluated the effect of sc-SWCNT chirality on the efficiency of sc-SWCNT/fullerene photovoltaic devices using both sc-SWCNTs and PCBM as light absorbers. The chirality of sc-SWCNTs was demonstrated to play key role in controlling the charge transfer efficiency at the interface, and the charge separation efficiency can be further enhanced by choosing the sc-SWCNTs with a higher conduction band level.

8.4.2 All-Carbon Planar Heterojunctions

Except for bulk heterojunctions, all-carbon planar heterojunctions have also been adopted, especially in polymer-free carbon-based photovoltaic devices, where a sharp atomic level planar interfaces is usually formed by depositing sc-SWCNTs and fullerenes in turn. Bindl et al. [23] demonstrated a NIR planar bilayer heterojunction photovoltaic device, in which highly purified sc-SWCNT film were employed as NIR optical absorbers and C_{60} was used as the electron acceptor. The PCE was found to be determined by the effective diffusion length of exciton migration in the sc-SWCNT films, and internal QE for exciton dissociation and charge collection approached 100% when the film thickness was smaller than diffusion length for the nanotube diameters <1.0 nm. As a result, the sc-SWCNT/C_{60} heterojunction photovoltaic device exhibited a PCE of 0.6% with a peak EQE >12% under NIR (1000–1365 nm) illumination with the density of 17 mW cm^{-2}. Furthermore, the efficiency of photocurrent generation through different optical excitation (E_{11}, $E_{11} + X$ and E_{22}) in the sc-SWCNTs was quantified in the carbon-based photovoltaic devices fabricated using nearly monochiral (7,5) sc-SWCNT/C_{60} bilayer heterojunctions [24]. The results suggested that the internal QEs generated by the optical excitation of E_{11}, E_{11} + X and E_{22} resonances were 85% ± 5%, 84% ± 14%, and 84% ± 7%, respectively. High internal QE indicated that free charge carriers can be efficiently generated from photons with the energy above E_{11} throughout the visible and NIR spectra using monochiral (7,5) sc-SWCNTs as photoabsorbers and C_{60} as acceptors. Consequently, a max EQE of 34% was achieved and a PCE of 7.1% with V_{OC} of 492 mV and FF of 62% was demonstrated under 1053 nm light illumination (100 mW cm^{-2}), as shown in Fig. 8.2. In addition, the EQE spectra and the photocurrent contributions of sc-SWCNTs and C_{60} can be quantitative predicted based on an optical model constructed by accurately measuring the wavelength-dependent optical constants of each layer [25].

The band offset between the LUMO of sc-SWCNTs and C_{60} plays an important role in determining the charge transfer at the interface. To promote the efficient charge transfer from sc-SWCNTs to C_{60}, (6,5)-enriched CoMoCAT sc-SWCNTs with large bandgap were used to fabricate sc-SWCNT/C_{60} photovoltaic devices by selectively dispersing with regioregular poly(3-dodecylthiophene-2,5-diyl) (rr-P3DDT) [26]. Compared with the 0.39 V Voc of device fabricated from HiPco sc-SWCNTs, the

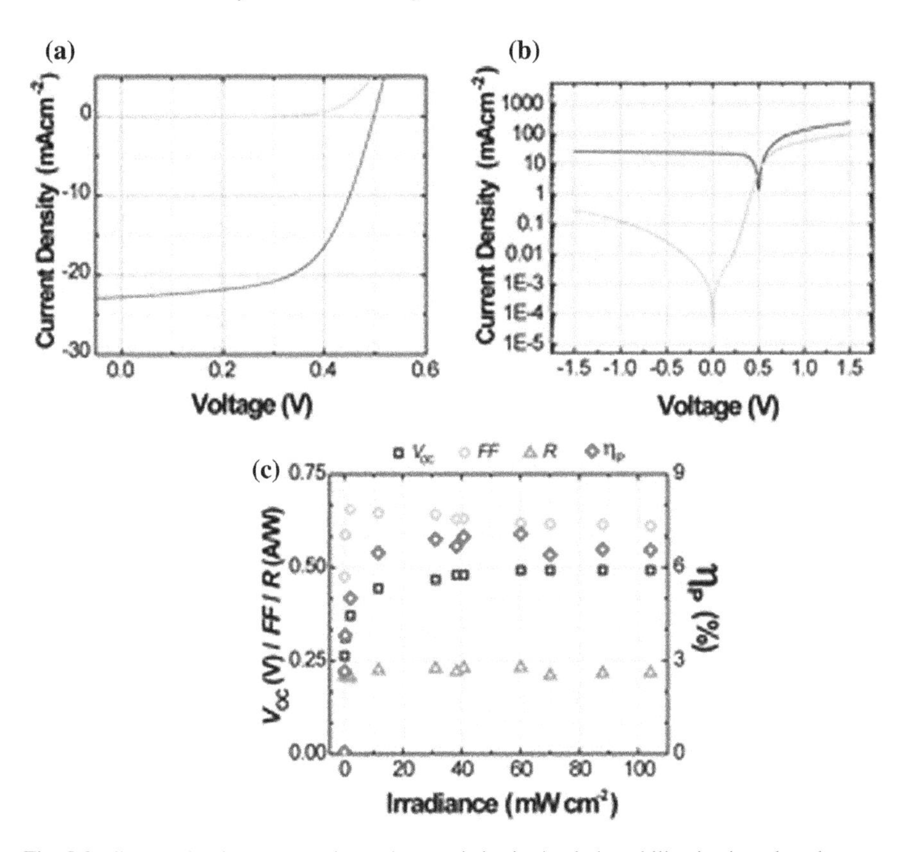

Fig. 8.2 Current density versus voltage characteristics in the dark and illumination plotted on **a** a linear and **b** log–linear scale. **c** Photovoltaic device parameters versus irradiance. Reproduced with permission [24]. Copyright 2013 American Chemical Society

carbon-based devices exhibited higher Voc of 0.44 V due to the smaller diameter and lower bandgap tubes in the sorted CoMoCAT sc-SWCNTs. On the other hand, the exciton diffusion length around 5 nm also limits the carrier extraction efficiency. Koleilat et al. [27] further demonstrated a sc-SWCNT/C_{60} vdW heterojunction active layer with enhanced carrier extraction efficiency, in which the photogenerated carriers could be effectively extracted from beyond the 5 nm limit within nanoporous sc-SWCNT film by C_{60} layer due to larger interfacial contacts. Meanwhile, highly doped C_{60} layer was adopted to increase the band offset between the LUMO levels of SWCNTs and C_{60}, promoting the efficient exciton separation at the interface. With a combination of minimizing the polymer content, tenfold improved performance have been achieved using 5 times thicker sc-SWCNT films compared to the flat bilayer structures (Fig. 8.3).

Although the abovementioned sc-SWCNTs with large bandgap has been demonstrated to fabricate sc-SWCNT/fullerene photovoltaic devices with fullerene or its derivatives, the V_{OC} of devices is still far lower than that of the bandgap of

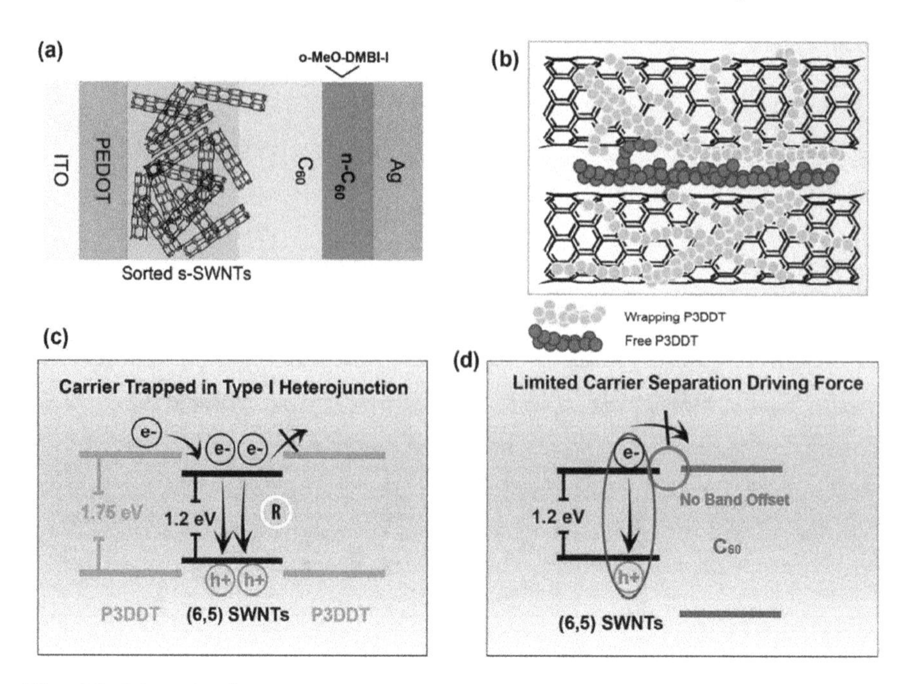

Fig. 8.3 Schematic diagrams of **a** sc-SWCNT/C_{60}/n-doped C_{60} structure and **b** s-SWNT surrounded by the sorting polymer P3DDT, **c–d** Two major carrier transport problems hindering sc-SWCNT contribution to the performance. Reproduced with permission [27]. Copyright 2016, American Chemical Society

sc-SWCNTs. Classen et al. [28] systematically investigated the V_{OC} and V_{OC} losses in the sc-SWCNT/fullerene photovoltaic devices, which was fabricated by spin-coating PFO-BPy wrapped (6,5) SWCNTs and $PC_{70}BM$ active layers, as shown in Fig. 8.4. Under AM 1.5 G spectrum at 100 mW cm^{-2}, the solar cell exhibits an overall PCE of 2.9% and a high V_{OC} of 0.59 V, which is a low loss compared with the bandgap of (6,5) SWCNTs. It is found that this low voltage losses were contributed to the lack of a measurable charge transfer state and the narrow absorption edge of SWCNTs. Moreover, the presence of other narrow-bandgap sc-SWCNTs limits the improvement of V_{OC}, the V_{OC} loss (Eg/q–V_{OC}) below 0.6 V would be achieved using pure single-chirality (6,5) SWCNTs.

To avoid high-bandgap polymers from blocking the direct contact between photogenerated carriers and the fullerene acceptor, Jain et al. [29] firstly demonstrated a polymer-free NIR photovoltaics without polymeric active or transport layers using highly purified sc-SWCNTs as the photo absorption layer and C_{60} as electron transport layer (Fig. 8.5). It was found that only a 20% another-chirality sc-SWCNT in weight would result in a more than 30 times decrease in the PCE due to the remarkably excitons trapping in the sc-SWCNT film. This study also

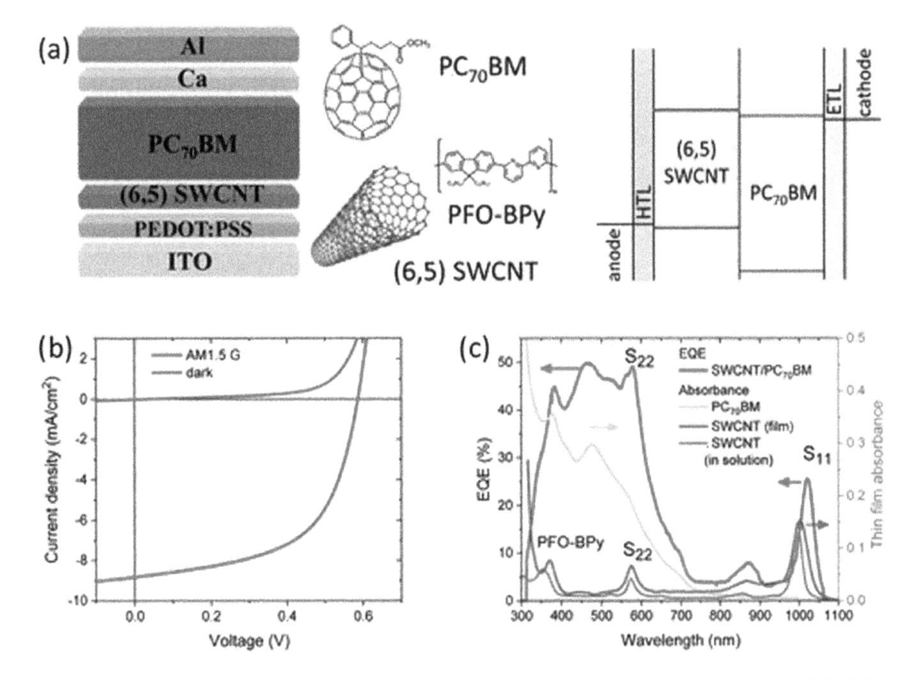

Fig. 8.4 **a** Device stack and schematic energy level diagram between (6,5) SWCNT and $PC_{70}BM$. **b** J-V curve and **c** EQE curve of the SWCNT/$PC_{70}BM$ solar cell. Reproduced with permission [28]. Copyright 2019, Wiley

provided several methods to improve the performance of devices, such as the thickness modulation of active layer, the architecture optimization of heterojunctions, sc-SWCNT alignment, and so on. Tung et al. [30] demonstrated a surfactant-free, water-processable all-carbon active layer for sc-SWCNT/C_{60} photovoltaic devices, in which pristine sc-SWCNTs, C_{60}, and graphene oxide (GO) were co-assembled in water and sc-SWCNT/C_{60}/reduced GO hybrids were then formed after thermal reduction. Using an additional C_{60} block layer, the photovoltaic device exhibited a PCE of 0.21% with a V_{oc} of 0.59 V. Shea et al. [31] further optimized the layer thickness and optical interference effects by using ultrathin highly mono-chiral (7,5) sc-SWCNT film (<5 nm). After the remove of excess PFO, it was demonstrated that the PCE of 0.95% could be achieved under AM1.5 G illumination with the max EQE of 43% at the nanotube bandgap (1055 nm). With the optimization of layer thicknesses by theoretical modeling, the distribution of the electric field within the bilayer heterojunctions can be matched to the light absorption by the (6,5)-enriched sc-SWCNTs through either their S_{11} or S_{22} absorption peaks. As a result, the sc-SWCNT/C_{60} solar cells with a high internal QE of 86% and a Voc of 0.44 V have been demonstrated using polymer-free (6,5)-enriched sc-SWCNTs [32].

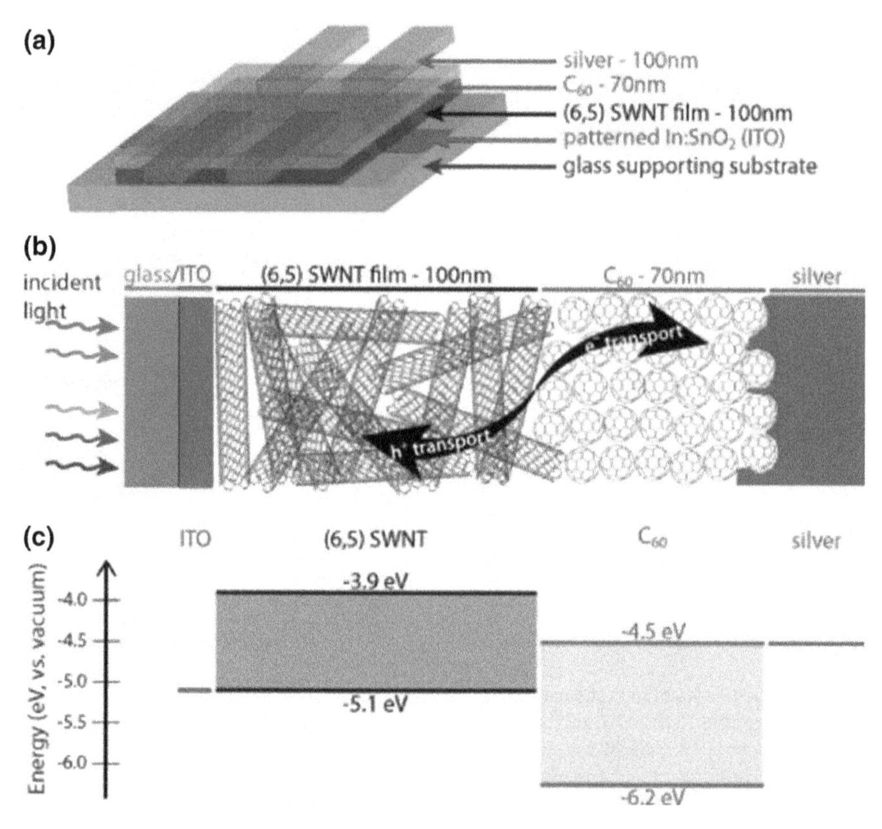

Fig. 8.5 Schematic diagram of **a** three-dimensional device architecture and **b** cross-sectional diagram. **c** Energy diagram of polymer-free carbon nanotube photovoltaic device. Reproduced with permission [29]. Copyright 2012, Wiley

8.5 All-Carbon Solar Cells

Although different all-carbon active layers have been demonstrated in carbon-based solar cells where sc-SWCNTs are used as light-absorbing material while fullerenes are used as the electron transport materials, the cathode and anode electrodes are usually typical metal (Al or Ag) and transparent oxide (ITO), respectively. No all-carbon solar cells composed entirely of carbon nanomaterials has been reported. Tung et al. [33] firstly proposed a conceptual, true all-carbon solar cell device (Fig. 8.6) on the basis of their previous research works. In the device model, graphene film or m-SWCNT network is used as transparent electrode to replace the ITO, and the active layer is composed of the mixture of fullerene/SWCNTs/rGO. The transportation of the photogenerated holes can be promoted by introducing a thin GO or GO/SWCNT film as interfacial layer between active layer and anode electrode if needed, while the photogenerated electrons are transported in the

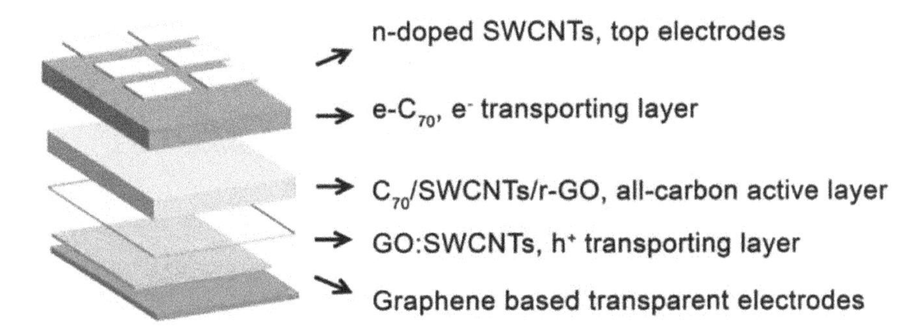

Fig. 8.6 Schematic diagram of a conceptual all-carbon solar cell. Reproduced with permission [33]. Copyright 2012, RSC

fullerenes layer. Finally, the solar cells can be fabricated by choosing n-doped SWCNTs or graphene as cathode electrode.

The first true all-carbon solar cells have been reported in 2013 by Bao group [34], in which both active layer and electrodes were composed of carbon-based materials. Specifically, the polymer sorted sc-SWCNTs film as light absorber formed p/n junctions with thermally evaporated C_{60} film, and rGO and n-doped SWCNT were used anode and cathode, respectively. Although the first all-carbon solar cell exhibited a PCE of $4.1 \times 10^{-3}\%$ under NIR illumination, the feasibility of carbon-based solar cells composed of all-carbon nanomaterials has been demonstrated. Obviously, many big challenges have to be overcome so that the PCE can be improved significantly. For example, the diameter and alignment of sc-SWCNTs need to be controlled and designed; the absorption peaks of sc-SWCNTs cannot overlap well with the solar spectrum; the doped SWCNT cathode should be fabricated directly on the top of C_{60} layer without damaging the underlying layer. All-carbon solar cells with PCE of >1% are expected to achieved after material and process optimization [34] (Fig. 8.7).

8.6 Summary and Outlook

In summary, the bandgap structure limits for the sc-SWCNT/fullerene hybrids have been first discussed, which provide an important guideline for designing the all-carbon nanohybrids in the solar cells. Then, the solar cells based all-carbon nanohybrid active layer have been highlighted, including pure sc-SWCNT/fullerene hybrids, polymer-wrapped hybrids, multi carbon nanomaterial hybrids, and so on. Finally, the concept and research progress about all-carbon solar cells composed of carbon active layer and carbon electrodes have been introduced as well as the corresponding challenges. Although the use of carbon nanostructures for all-carbon photovoltaic devices has been demonstrated, there are still many aspects to be solved or well-designed before all-carbon solar cells become a reality.

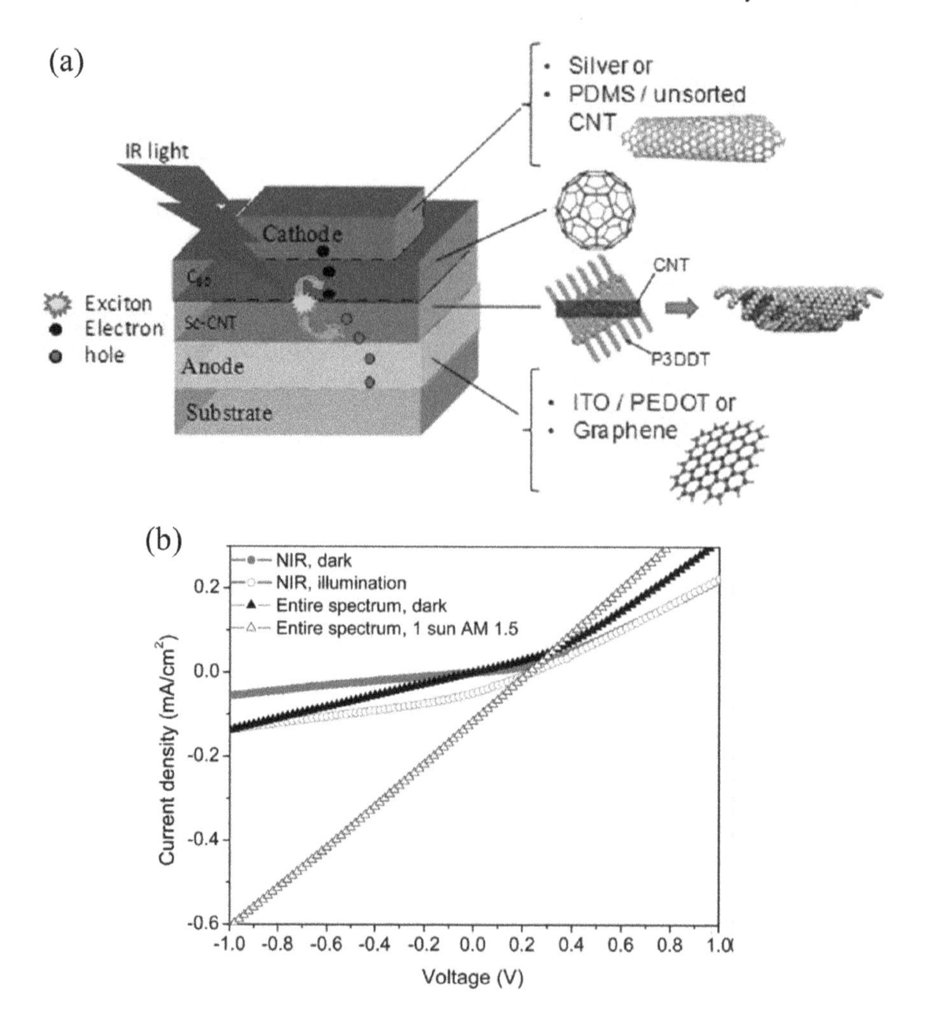

Fig. 8.7 **a** Structure diagram of sc-SWCNT based photovoltaic cell. **b** J-V curves of all-carbon solar cell. Reproduced with permission [34]. Copyright 2012 American Chemical Society

The first and most important thing is to increase the light absorption of all-carbon active layer, which has limited significantly the PCE increase of all-carbon solar cells. The exciton diffusion length perpendicular to the sc-SWCNT films is only a few nanometers, resulting in a low light absorption of the ultrathin sc-SWCNT films. For a single active layer composed of sc-SWCNT layer and fullerene layer, the increase of the thickness of sc-SWCNT films has to balance with the transport efficiency of the photogenerated carriers perpendicular to the films. In principle, the alternative deposition strategy of fullerene layer and sc-SWCNT layer can be used to effectively solve the abovementioned problems. So the next important step is to prepare and deposit the ultrathin sc-SWCNT films with uniform thickness. In addition, the built-in electric field between fullerenes and sc-SWCNTs needs to be enhanced by controlling their Fermi levels, promoting effectively the charge separation of photogenerated carriers at the interface. Except for the laminated structure, the planar device structures are also expected to improve the PCE of all-carbon solar cells. Fullerenes contact one end of the aligned sc-SWCNTs, and the other end of sc-SWCNTs directly bridge the anode electrode. Consequently, the photogenerated carriers in sc-SWCNTs transport along the axial direction and can be separated efficiently at the interface between fullerenes and the end of sc-SWCNTs.

Another important thing is to obtain single-chirality sc-SWCNTs with ultrahigh purity in large scale. On the one hand, the electronic bands of sc-SWCNTs are closely related to the chiralities, only a few sc-SWCNTs with specific band structures are suitable for the all-carbon active layer with fullerenes due to the bandgap alignments. On the other hand, the existence of the m-SWCNTs and sc-SWCNTs with other chiralities results in a high recombination probability of the photogenerated carriers in the sc-SWCNT film. Nowadays, the selective growth or highly efficient separation of single-chirality sc-SWCNTs is still a big challenge, and only a few sc-SWCNTs with specific chirality can be obtained and the purity is still low. Therefore, more researches on the enrichment of single-chirality sc-SWCNTs need to be further strengthened in the future.

As a top carbon electrode, the high transparency of the SWCNT film with high conductivity is very important so that as much light as possible can pass through, However, it is not usually achieved simultaneously due to the process challenges of ultrathin SWCNT film, more efforts are still needed to make to directly deposit the top carbon electrode on the all-carbon active layer without damage. Another important thing is to regulate the workfunction of SWCNT film, so that it can form good Ohmic contact with fullerene layer. Similarly, the workfunction of mono- or few-layer graphene also needs to be tailored according to the chirality of sc-SWCNTs in the active layer.

After the design and optimization of the abovementioned carbon nanomaterials and device structures, all-carbon solar cells composed of carbon-based electrode and active layer can be fabricated, exhibiting higher power conversion efficiency toward NIR spectrum. In order to absorb efficiently the full spectrum of sunlight, the strategy of dual-junction or multijunction solar cells can be adopted by stacking III–V solar cells as top sub-cell on the top of all-carbon solar cells, in which III–V solar cells mainly respond to UV and visible light. We believe that higher power conversion efficiency can be obtained by optimizing the tunneling junction and current matching design.

References

1. Jariwala D, Sangwan VK, Lauhon LJ, Marks TJ, Hersam MC (2013). Carbon nanomaterials for electronics, optoelectronics, photovoltaics, and sensing. Chem Soc Rev, 42(7): 2824–2860.
2. Liu Y, Wang S, Liu H, Peng LM (2017). Carbon nanotube-based three-dimensional monolithic optoelectronic integrated system. Nat Commun 8(1): 1–8.
3. Lin S, Lu Y, Xu J, Feng S, Li J (2017). High performance graphene/semiconductor van der Waals heterostructure optoelectronic devices. Nano Energy, 40: 122–148.
4. Di Bartolomeo A (2016). Graphene Schottky diodes: An experimental review of the rectifying graphene/semiconductor heterojunction. Phys Rep, 606: 1–58.
5. Martín N (2017). Carbon nanoforms for photovoltaics: Myth or Reality? Adv Energy Mater, 7 (10): 1601102.
6. Cai BF, Su YJ, Tao ZJ, Hu J, Zou C, Yang Z, Zhang YF (2018). Highly sensitive broadband single-walled carbon nanotube photodetectors enhanced by separated graphene nanosheets. Adv Optical Mater, 6(23): 1800791.
7. Avouris P, Freitag M, Perebeinos V (2008). Carbon-nanotube photonics and optoelectronics. Nat Photonics, 2(6): 341–350.
8. Pfohl M, Glaser K, Graf A, Mertens A, Flavel BS (2016). Probing the diameter limit of single walled carbon nanotubes in SWCNT: fullerene solar cells. Adv Energy Mater, 6(21): 1600890.
9. Mehlenbacher RD, Wu MY, Grechko M, Laaser JE, Arnold MS, Zanni MT (2013). Photoexcitation dynamics of coupled semiconducting carbon nanotube thin films. Nano Lett, 13(4): 1495–1501.
10. Grechko M, Ye Y, Mehlenbacher RD, McDonough TJ, Wu MY, Jacobberger RM, Arnold MS, Zanni MT (2014). Diffusion-assisted photoexcitation transfer in coupled semiconducting carbon nanotube thin films. ACS Nano, 8(6): 5383–5394.
11. Kafle TR, Wang T, Kattel B, Liu Q, Gong Y, Wu J, Chan WL (2016). Hot exciton relaxation and exciton trapping in single-walled carbon nanotube thin films. J Phys Chem C, 120: 24482–24490.
12. Ferguson AJ, Dowgiallo AM, Bindl DJ, Mistry KS, Reid OG, Kopidakis N, Arnold MS, Blackburn JL (2015). Trap-limited carrier recombination in single-walled carbon nanotube heterojunctions with fullerene acceptor layers. Phys Rev B, 91(24): 245311.
13. Dowgiallo AM, Mistry KS, Johnson JC, Blackburn JL (2014). Ultrafast spectroscopic signature of charge transfer between single-walled carbon nanotubes and C60. ACS Nano, 8 (8): 8573–8581.
14. Bindl DJ, Ferguson AJ, Wu MY, Kopidakis N, Blackburn JL, Arnold MS (2013). Free carrier generation and recombination in polymer-wrapped semiconducting carbon nanotube films and heterojunctions. J Phys Chem Lett, 4(21), 3550–3559.
15. Dowgiallo AM, Mistry KS, Johnson JC, Reid OG, Blackburn JL (2016). Probing exciton diffusion and dissociation in single-walled carbon nanotube-C_{60} heterojunctions. J Phys Chem Lett, 7(10): 1794–1799.
16. Tune DD, Shapter JG (2013). The potential sunlight harvesting efficiency of carbon nanotube solar cells. Energy Environ Sci, 6, 2572–2577.
17. Bindl DJ, Brewer AS, Arnold MS (2011). Semiconducting carbon nanotube/fullerene blended heterojunctions for photovoltaic near-infrared photon harvesting. Nano Res, 4(11): 1174–1179.
18. M Gong, TA Shastry, Y Xie, M Bernardi, D Jasion (2014). Polychiral semiconducting carbon nanotube-fullerene solar cells. Nano Lett, 14(9): 5308–5314.
19. Gong M, Shastry TA, Cui Q, Kohlmeyer RR, Luck KA, Rowberg A, Marks TJ, Durstock MF, Zhao H, Hersam MC, Ren S (2015). Understanding charge transfer in carbon nanotube-fullerene bulk heterojunctions. ACS Appl Mater Interfaces, 7(13): 7428–7435.

20. Shastry TA, Clark SC, Rowberg AJ, Luck KA, Chen KS, Marks TJ, Hersam MC (2016). Enhanced uniformity and area scaling in carbon nanotube-fullerene bulk-heterojunction solar cells enabled by solvent additives. Adv Energy Mater, 6(2): 1501466.
21. Bernardi M, Lohrman J, Kumar PV, Kirkeminde A, Ferralis N, Grossman JC, Ren S (2012) Nanocarbon-based photovoltaics. ACS Nano, 6(10), 8896–8903.
22. Isborn CM, Tang C, Martini A, Johnson ER, Otero-de-la-Roza A, Tung VC (2013). Carbon nanotube chirality determines efficiency of electron transfer to fullerene in all-carbon photovoltaics. J Phys Chem Lett, 4(17): 2914–2918.
23. Bindl DJ, Wu MY, Prehn FC, Arnold MS (2011). Efficiently harvesting excitons from electronic type-controlled semiconducting carbon nanotube films. Nano Lett, 11(2): 455–460.
24. Bindl DJ, Arnold MS (2013). Efficient exciton relaxation and charge generation in nearly monochiral (7,5) carbon nanotube/C_{60} thin-film photovoltaics. J Phys Chem C, 117(5): 2390–2395.
25. Guillot SL, Mistry KS, Avery AD, Richard J, Dowgiallo AM, Ndione PF, van de Lagemaat J, Reese MO, Blackburn JL (2015). Precision printing and optical modeling of ultrathin SWCNT/C_{60} heterojunction solar cells. Nanoscale 7(15), 6556–6566.
26. Wang H, Koleilat GI, Liu P, Jiménez-Osés G, Lai YC, Vosgueritchian M, Fang Y, Park S, Houk KN, Bao ZN (2014). High-yield sorting of small-diameter carbon nanotubes for solar cells and transistors. ACS Nano, 8: 2609–2617.
27. Koleilat GI, Vosgueritchian M, Lei T, Zhou Y, Lin DW, Lissel F, Lin P, To JWF, Xie T, England K, Zhang Y, Bao ZN (2016). Surpassing the exciton diffusion limit in single-walled carbon nanotube sensitized solar cells. ACS Nano, 10(12): 11258–11265.
28. Classen A, Einsiedler L, Heumueller T, Graf A, Brohmann M, Berger F, Kahmann S, Richter M, Matt GJ, Forberich K, Zaumseil J (2019). Absence of charge transfer state enables very low V_{OC} losses in SWCNT: fullerene solar cells. Adv Energy Mater, 9(1), 1801913.
29. Jain RM, Howden R, Tvrdy K, Shimizu S, Hilmer AJ, Mcnicholas TP, Gleason KK, Strano MS (2012). Polymer-free near-infrared photovoltaics with single chirality (6,5) semiconducting carbon nanotube active layers. Adv Mater, 24(32): 4436–4439.
30. Tung VC, Huang JH, Tevis I, Kim F, Kim J, Chu CW, Stupp SI, Huang JX (2011). Surfactant-free water–processable photoconductive all-carbon composite. J Am Chem Soc, 133(13): 4940–4947.
31. Shea MJ, Arnold MS (2013). 1% solar cells derived from ultrathin carbon nanotube photoabsorbing films. Appl Phys Lett, 102: 243101.
32. Pfohl M, Glaser K, Ludwig J, Tune DD, Dehm S, Kayser C, Colsmann A, Krupke R, Flavel BS (2016). Performance enhancement of polymer-free carbon nanotube solar cells via transfer matrix modeling. Adv Energy Mater, 6(1): 1501345.
33. Tung VC, Huang JH, Kim J, Smith AJ, Chu CW, Huang JX (2012). Towards solution processed all-carbon solar cells: a perspective. Energy Environ Sci, 5(7): 7810–7818.
34. Ramuz MP, Vosgueritchian M, Wei P, Wang CG, Gao YL, Wu YP, Chen YS, Bao ZN (2012). Evaluation of solution-processable carbon-based electrodes for all-carbon solar cells. ACS Nano, 6(11): 10384–10395.

Index

A

Absorption coefficient, 13, 14, 37, 39, 92, 94, 96, 102, 113, 131, 140, 149, 164, 171, 174
All-carbon p/n junctions, 11
All-carbon solar cell, 171–173, 180–183
All-carbon vdW heterostructures, 132, 134, 137, 138, 141, 144
Alloy catalysts, 50, 81
Antireflection layer, 156, 161, 163
Arc-discharge, 5, 41, 139
Arc plasma, 43
Armchair tubes, 27–29
Arrays, 8, 10, 11, 18, 20, 48, 50–52, 97–101, 105, 112, 113, 135, 137, 144
Aspect ratio, 16, 31
Assignment, 38, 65, 69, 70, 76, 81, 84, 85
Atom configuration, 91
Atomic-level interface, 102
Atomic structures, 2, 4, 27, 31, 65, 76, 79, 85
Atomic thickness, 6, 8

B

Ballistic transport, 8, 33, 34, 149
Band alignment, 107, 108, 135
Band gap, 3, 31, 65, 103, 105, 107, 108, 110, 120, 131
Band structures, 10, 11, 27, 31, 32, 37, 38, 122, 138, 143, 183
Black Phosphorus (BP), 12, 116, 120, 121, 163
Bolometric effects, 11, 92, 93, 96
Built-in electric field, 12, 93, 97, 104, 108, 138, 139, 151, 156, 160, 163, 183

C

Carbon allotropes, 1–3, 7, 11, 131, 132, 171
Carbon atoms, 2–4, 6, 7, 18, 27, 29, 30, 41, 45, 71, 75, 85, 109, 138, 171
Carbon-based optoelectronic nanodevices, v
Carbon nanomaterials, 1, 2, 4–6, 16, 20, 73, 91, 121, 122, 131, 132, 143, 171, 180, 181, 183
Carbon Nanotubes (CNTs), 1–3, 5, 6, 8, 15, 16, 20, 27, 41, 65, 66, 75, 91, 97, 100, 102, 103, 106, 131, 149, 152, 153, 155, 156, 166, 167, 171, 180
Carbon Quantum Dots (CQDs), 4, 137
Carbon sources, 41, 46–52, 57
Carrier mobility, 3, 6, 9, 10, 13, 14, 20, 36, 91, 92, 94–96, 102, 103, 107, 113, 116, 117, 120, 134, 137, 138, 149, 171, 174
Catalyst precursor, 46, 47
Catalytic growth, 55, 56
Catalytic substrates, 53
Characterizations, 41, 54, 65, 66, 75, 76, 85
Charge transfer, 13, 15, 17–19, 66, 67, 73–75, 103, 106–108, 116, 117, 122, 132, 134, 137–139, 160, 163, 167, 176, 178
Chemical doping, 10, 12, 14, 67, 75, 97, 122, 150, 154–156, 158, 160, 167
Chemical exfoliation, 7, 53
Chemical Vapor Deposition (CVD), 7, 41, 46–53, 55–58, 107, 116, 133, 158, 159
Chiral angle, 5, 27, 28, 48, 65, 68, 76, 77, 84, 85
Chiral-depended bandgap, 10
Chiral indices (n, m), 5, 27, 28, 33, 38, 76, 77

Y. Su, *High-Performance Carbon-Based Optoelectronic Nanodevices*, Springer Series in Materials Science 319,
https://doi.org/10.1007/978-981-16-5497-8

Chirality, 5, 29, 35, 48–51, 58, 65–68, 81–83, 143, 149, 157, 167, 171, 173, 176, 178, 183
Chirality control, 51
Chirality distribution, 43, 47–49, 65, 76, 79, 84
Chirality-selective synthesis, 19, 20, 46, 58
Chiral tubes, 27, 28
Conducting type, 149
Crystal orientation, 58
Curvature effect, 38

D

Dark current, 91, 93, 94, 103, 109, 121, 141–144, 164
Decomposition, 46, 47, 50, 56
Density Of States (DOS), 3, 10, 31–33, 67, 73, 95
Detectivity, 10, 12, 93, 94, 98–101, 103–106, 108, 111, 113–119, 121, 135–138, 140, 141
Diameter, 3–6, 8, 10, 13, 19, 27, 28, 33, 36, 37, 42, 43, 45–49, 65–69, 75, 76, 79, 81, 84, 85, 97, 103, 107, 131, 139, 149, 150, 157, 159, 160, 171, 173, 174, 176, 177, 181
Diameter distribution, 43, 45–47, 49, 52, 58, 81, 96, 100
Dielectric substrates, 7, 56
Dielectrophoresis, 98, 142
Diffraction spots, 76, 79
Dirac cones, 122
Dirac electrons, 39
Direct bandgap, 6, 14, 92, 113, 117, 120, 164, 171
Direct Current (DC), 41, 42
Double-Walled Carbon Nanotubes (DWCNTs), 106, 153, 162, 167

E

Electrochemical Energy Storage (EES), 16–18
Electron coupling effect, 12, 13, 121, 122, 131, 132, 143
Electron Diffraction (ED), 65, 76, 79
Electron-hole pairs, 37, 39, 82, 104, 134, 151, 152, 157
Electronics, 1–5, 7–10, 18–20, 27, 29, 31, 33, 41, 53, 58, 65–67, 75, 79, 82, 83, 85, 91, 92, 96, 100, 103, 107, 109, 132, 139, 160, 165, 167, 171, 172, 183
Energy storage, 1, 4, 20
Epitaxial growth, 7, 53, 112
Exciton, 10, 13, 37, 82, 92, 96–98, 105, 106, 121, 122, 134, 135, 138, 139, 141, 143, 144, 171–178, 182
Excition binding energy, 3, 37, 101
Exciton dissociation, 139, 172, 176
External Quantum Efficiency (EQE), 94, 95, 104, 106, 110, 116, 117, 136, 152, 176, 179

F

Fast Fourier Transformation (FFT) patterns, 75
Fermi level, 3, 31–34, 117, 122, 138, 139, 143, 149, 151–153, 155, 156, 160, 162, 166, 183
Fermi velocity, 30, 36, 109
Field-effect Transistors (FETs), 14, 15
Fill Factor (FF), 13, 152–154, 156, 158–160, 162, 166, 175, 176
Floating catalyst CVD methods, 47
Fluorescence, 3, 19, 38, 83, 84, 106, 131
Fullerene, 1–4, 6, 11, 13, 41, 131, 132, 134, 137, 171–178, 180–183

G

Gain (G), 95, 135
Grain domains, 79
Graphdiyne, 2, 3, 7
Graphene, 1–20, 27–33, 36, 38, 39, 41, 53–58, 65, 70–76, 78–80, 84, 85, 92, 109–122, 131–144, 154, 171, 181, 183
Graphene films, 7, 14, 18, 53–58, 133, 139, 180
Graphene glass, 56, 57
Graphene Oxide (GO), 3, 7, 19, 113, 175, 179, 180
Graphene Quantum Dots (GQDs), 3, 4, 137, 138
Growth mechanism, 20, 41, 44, 50

H

Heterojunctions, 10–15, 73, 74, 85, 96, 101–104, 106–109, 112–117, 122, 131, 132, 135, 138, 139, 142–144, 149–157, 159–167, 171–173, 175–177, 179
High-pressure Carbon Monoxide (HiPco), 47, 176
High-temperature plasma, 41, 42
Hole transport, 13, 162, 163
Hybridization, 2, 3, 31, 91, 131, 132, 138, 149, 171
Hydrocarbon gases, 47

I

Incident photon, 91, 94, 95, 151, 152
Interband optical transitions, 38, 65, 79
Interface passivation, 14, 104

Interfaces, 6, 11, 12, 14, 18, 34, 37, 74, 84, 92, 103, 104, 107–110, 113, 114, 116, 117, 121, 122, 131, 132, 134–140, 143, 144, 149–152, 155, 157, 158, 160, 163, 165, 166, 172, 173, 176, 177, 183
Interfacial interaction, 107

K
Kataura plots, 33, 68, 69

L
Laser ablation method, 45
Lifetime, 3, 12, 37, 52, 93–95, 110, 112, 137, 172
Light-mater interaction, 13, 121, 122
Localized surface plasmon resonance, 12, 114
Lorentzian peak, 70, 71
Low-cost solution processing, 13

M
Magnetic field, 15, 32, 43, 44
Main group metal chalcogenides, 12, 116, 118
Mechanical exfoliation, 7, 53, 116
Metallic, 1, 3–6, 10, 13, 14, 32–36, 47, 49–51, 66, 68, 81, 103, 122, 131, 149–153, 157, 171
Monolayers, 5–7, 11, 12, 27, 31, 38, 53, 54, 57, 58, 65, 70, 72, 75, 78–80, 84, 92, 107, 110–117, 122, 133, 140, 142
Multiphoton Absorption (MPA), 92, 96
Multiple Exciton Generation (MEG), 96, 106, 113
Multi-Walled Carbon Nanotubes (MWCNTs), 6, 15, 18–20, 106, 150, 151, 153–157, 167

N
Nanodevices, 2, 6, 7, 10, 12, 14, 15, 20, 91, 92, 97, 131, 135, 143, 171
Nanosheets, 7, 11, 12, 17, 107, 108, 113, 116, 117, 120, 121, 134, 137, 141, 163
Nanostructures, 4, 7, 78, 108, 110, 122, 131, 181
Near Infrared (NIR), 10, 11, 13, 19, 80, 82, 83, 91, 96, 98, 99, 101, 103, 106, 107, 113–116, 118, 120, 137, 139, 140, 172–176, 178, 181, 183
Nonmetal substrates, 55
Nucleation process, 51

O
Ohmic contact, 34, 97, 105, 133, 164, 167, 183
One-Dimensional (1D), 3, 5, 65, 131, 149
Optical absorption, 5, 10, 12, 14, 37–39, 66, 92, 96, 101, 107, 110, 112–114, 135, 164, 171, 172
Optical properties, 4, 7, 13, 19, 27, 37–39, 58, 109, 132, 135, 166
Optoelectronic, 2, 3, 6, 7, 10, 12, 18, 20, 37, 65, 91, 92, 97, 107, 109, 111, 117, 120, 131, 132, 171

P
Phonons, 34–36, 39, 66, 67, 74
Photoactive materials, 11, 91, 103
Photoconductors, 10, 93, 95, 112
Photodetectors, 4, 10–12
Photodiode, 10, 12, 13, 93, 97, 98, 103, 107, 110, 111, 134, 144
Photoelectric properties, 10, 13, 101–103, 106, 112, 114, 116, 139, 149, 165, 172
Photogating effect, 12, 92, 93, 110, 111, 114, 140
Photoluminescence (PL), 4, 10, 38, 65, 82
Photoluminescence Excitation (PLE), 38, 83, 84
Phototransistor, 10–12, 93, 106–108, 110, 111, 114–117, 121, 134–137, 139
Photovoltage, 92, 94, 100, 101, 103, 112, 175
Photovoltaic, 2, 4, 13, 37, 92, 93, 97, 98, 103, 104, 113, 117, 120, 134, 149, 152, 155, 161–167, 171, 172, 174–182
Plasma parameters, 43
PLE mapping, 83
Polycrystalline, 7, 53, 54, 65, 79, 80
Powders, 48, 81
Power Conversion Efficiency (PCE), 13, 14, 97, 149, 150, 152–163, 166, 167, 175, 176, 178, 179, 181–183
Preferential growth, 49, 50, 52, 53, 122
Purity, 20, 42, 43, 45, 50, 52, 58, 65, 81, 149, 183

Q
Quantum confinement effects, 4, 117
Quantum Dot (QD), 4, 12, 105, 113, 131, 132
Quantum size effect, 3, 105, 113

R
Radial Breathing Mode (RBM), 66–70
Raman scattering process, 73
Raman shift, 67, 73, 113, 139
Raman spectra, 44, 66, 67, 69, 71, 73–75, 79
Recombination, 10, 11, 13, 19, 38, 39, 82, 93, 95, 96, 108, 111, 113, 121, 122, 135, 137, 139, 144, 149, 153, 163, 165, 167, 172, 183

Reduced graphene oxide (rGO), 7, 16, 17, 19, 20, 113, 175, 180, 181
Responsivity, 10–12, 91, 93–95, 97–101, 103–106, 108, 110–121, 134–142

S
Saturable absorption, 37, 39
Scanning Electron Microscopy (SEM), 52, 98, 109, 111, 133
Scanning Tunneling Microscopy (STM), 85
Schottky barrier, 15, 93, 96, 117, 121, 163
Schottky contacts, 10
Selected-Area Electron Diffraction (SAED), 79, 80
Selective-synthesis, 10
Selectivity, 15, 16, 19, 51, 110
Semiconducting, 1, 3–6, 13, 32, 44, 48, 65, 68, 81, 92, 94, 102, 105–108, 113, 117, 131, 149, 150, 153, 157, 171
Sensitivity, 10, 15, 16, 19, 83, 100–102, 112, 122, 138, 144
Sensor, 7, 14–16, 19, 20, 73
Series resistance, 13, 160
Short-circuit Current Density (JSC), 13, 152, 153, 155, 158–160, 162, 165, 166, 175
Shunt resistance, 152
Single-chirality, 5, 12, 49–51, 121, 122, 143, 178, 183
Single crystal, 7, 50, 53, 54, 58
Single walled carbon nanotubes, 1, 41, 91, 131, 171
Solar cells, 4, 6, 13, 14, 103, 104, 106, 112, 149–167, 171–175, 178–183
Specific surface area, 6, 14, 15, 17, 19, 27, 48, 105
Structural characterizations, 65
Structural defects, 11, 34, 35, 43, 65, 66, 75, 107, 113, 144
Structure-specific growth, 49
Supercapacitors, 6, 16, 17
Surface states, 15, 16, 27, 120, 167
SWCNT film, 13, 18, 100–104, 157–162, 165, 183

T
Temperature Coefficient of Resistance (TCR), 93, 100
Thermal conductivity, 1, 4, 6, 7, 18, 42, 92, 171
Thermal detectors, 10, 11, 93, 96, 100
Thermal management, 6, 18
Thermoelectric effects, 100
Transition Metal Dichalcogenides (TMDs), 12, 116–118
Transmission Electron Microscopy (TEM), 5, 41, 75, 76, 78, 80
Two-Dimensional (2D), 1, 4, 6, 131, 132, 171

U
Ultrafast carrier dynamics, 92, 110
UV-vis-NIR absorption spectra, 65, 79, 80

V
Van der Waals (vdW) heterojunctions, 11, 107, 131
Van Hove singularities, 3, 6, 10, 32, 66, 96, 122, 172

W
Wafer-scale growth, 58
Waveguide, 110
Work function, 8, 97, 119, 151, 152, 156, 160

Y
Yield, 11, 41–43, 45, 47–49, 52, 58, 79, 134, 172

Z
Zero-bandgap, 3, 65
Zero-Dimensional (0D), 4, 131
Zigzag tubes, 27, 85

Lightning Source UK Ltd.
Milton Keynes UK
UKHW022357210922
409193UK00002B/16